Traveling at the Speed of Light

The Possibility of Light Speed

By Mark F. Dennis

This text has been drawn from my previous book, "A Challenge to Einstein". The focus of this book is to look at what light is, assess light speed, ponder whether or not the speed of light is attainable and evaluate the possibility that someday the speed of light may be surpassed.

Reading the book "The Speed of Light, Reviewing the History of c" prior to reading this book is highly recommended. The theories represented in this book cannot be thoroughly appreciated without the review of the history of "c".

"In short, let us assume that the simple law of the constancy of the speed of light c (in a vacuum) is justifiably believed by the child at school. Who would have imagined that this simple law has plunged the conscientiously thoughtful physicist into the greatest of intellectual difficulties?"

(Albert Einstein)

Relativity by Einstein, pg. 18

Table of Contents

Table of Figures

Traveling at the Speed of Light

The Possibility of Light Speed

The Theory
of Relativity

Preface

Einstein's Special Theory of Relativity was developed by Einstein after the failure of Michelson and Morley to detect the "luminiferous ether". The theory of the Constancy of the Speed of Light grew out of this study and was designed to solve the age-old riddle of why measurements of the speed of light did not show any variation with Earth's movement through space. The speed of light therefore had to be a constant that did not change with the motion of the earth nor any other object no matter what its velocity was. The aether that light was believed to move through also would no longer be necessary, neither would the idea of "stationary space".

As Einstein would put it:

"Examples of this sort, together with the unsuccessful attempts to discover any motion of the earth relatively to the "light medium," suggest that the phenomena of electrodynamics as well as of mechanics possess no properties corresponding to the idea of absolute rest.... Light is always propagated in empty space with a definite velocity of c which is independent of the state of motion of the emitting (or detecting) body. These two postulates suffice for the attainment of a simple and consistent theory of the electrodynamics of moving bodies based on Maxwell's theory for stationary bodies. The introduction of the "luminiferous ether" will prove to be superfluous inasmuch as the view here to be developed will not require an "absolute stationary space" provided with special properties, nor assign a velocity vector to a point in space in which electromagnetic processes take place.

(Einstein, 1905, pg. 1)

Einstein would also say:

"Who would have imagined that this simple law has plunged the conscientiously thoughtful physicist into the greatest of intellectual difficulties?"

(Einstein, 1952, pg. 18)

The Special Theory of Relativity would answer the long sought after questions of the past but introduce problems of its own. The standard physical equations for moving objects would no longer hold true. The velocity of a baseball thrown from a moving vehicle would normally be calculated as the vehicles velocity added to the velocity at which the baseball was thrown or $v1 + v2 = v3$. A baseball thrown at 50 miles per hour (mph) from a vehicle driven at 100 mph would be traveling at 150 mph relative to the ground and the person catching the ball. The speed of light is however always the same no matter what the velocity of the emitter or detector. Therefore, using the above equation would change and be written as, $v1 + v2 = v1$ or using the letter c for the speed of light, $c + v1 = c$ and $c - v1 = c$. This contradicts normal thinking, a paradox of sorts, creating some "difficulties". In order to explain this contradiction, Albert Einstein developed formulas to show how that movement of an object warps space and time and that space and time were really one entity, "space-time". He further explained how that acceleration and gravity where truly one in the same with his "Equivalence Principle", and that both also warped space and time.

The velocity of light was fixed; nothing could go faster than the speed of light. He would show through his relativistic mass formula that as an object would accelerate to faster and faster velocities that an objects mass would start to increase exponentially with velocity as energy would be turned into mass, founded on the famous formula $E = MC^2$. Mass would be an extension of energy as $M = C^2/E$. Nothing could go faster than light. This then would compound things to a certain degree. How could electrons and photons then travel at or close to the speed of light? The answer was to make these components massless particles.

As we can see, there is an interesting dialogue going on here. In order to answer one problem, a new problem is created. Are these theories truly the way things are or are they really only theories? Do we dare question or challenge them? As we review these theories, let's ask some basic questions of our own. What truly is energy? How is energy propagated through empty space? Is space really empty? What is attraction from a distance (gravity and electrostatic charge) and how can these forces operate across the vast emptiness of space or even from across the room? These were the questions that drove the genius of Albert Einstein to question the scientific theories that were in place in the early twentieth century. Now, almost 100 years later, Einstein's theories have become the cornerstone of quantum mechanics and theoretical physics. Today in the early twenty-first century, is it not time to question these old theories again?

The core question of this book and what we really want to know is, "is the speed of light attainable? Could someday the speed of light be surpassed? Will this affect time? Can we travel in time? Understanding the questions we are asking takes a basic knowledge of the subjects of Quantum Mechanics, General Relativity and Special Relativity. I would recommend reading "The Speed of Light, Reviewing the History of c", as a prerequisite before getting deeply entrenched into the subject at hand.

Review of History

Mankind has always wondered about the universe. Where did we come from and what are we made of? The questions about light, its composition and how it works has always intrigued us, even back to the 300 BCE's, in the days of Plato, Aristotle and Socrates (Dennis, pg. 1). In those days, light was considered the fifth element; the "luminiferous aether", the stuff stars were made of. It wasn't until Galileo in the early 1600's that the question of light's speed was tested with the use of covered lanterns. All that he could determine at that time was that the speed of light was at least 10 times faster than the speed of sound (Dennis, pg. 2). Not too much later in the late 1600's however, Ole Rømer determined through the study of Jupiter's moons that the speed of light was be 299,792 km/sec or 186,290 miles/sec, becoming the cornerstone of quantum physics, the study of atomic matter (Dennis, pg. 4).

Various theories of what light consisted of, slowly developed, including René Descartes undulation theory, otherwise known as wave theory in the early 1600's and Sir Isaac Newton's corpuscular theory or particle theory in the early 1700's (Dennis, pg. 5). Newton's particle theory however did not stand very long as Thomas Young in the late 1700's used the "Double Slit" experiment where light created interference patterns, proving that light was composed of waves (Dennis, pg. 8). Studies showed that light was made of both longitudinal waves and transverse waves complicating the aether theory of the day. This suggested that the aether, the medium that light traveled through, must be created of a solid substance, which would resist the motions of the planets and the stars and did not seem to make sense (Dennis, pg. 6).

In the 1700's, James Bradley became the first to prove the Copernican heliocentric view of the universe. It became evident through his studies that the earth moved very rapidly though space in its orbit around the sun. These studies of star aberration, or the apparent shift in a stars position, showed different displacements of the stars over time in the northern region around the North Star and south in the equatorial region (Dennis, pg. 9). This lead to his calculation of the speed of light which was very close to Ole Rømer's calculations 100 years earlier. This knowledge was reinforced by Fizeau's and Foucault's experiments showing lights speed within a couple thousand miles per second (Dennis, pg. 16).

Later on in the late 1700's François Arago tried to show that light traveled at different speeds due to gravitational forces of stars and the movement of the stars. His experiment, using a prism and a telescope was set up to detect these various speeds using Newton's refraction theory. The test failed; at least in showing that there was no variation in the refraction of light. This experiment added confusion to the mix by indicating that the speed of light from large stars and small stars, and that of moving stars in relation to the movement of the earth was the all same (Dennis, pg. 14).

The aether or medium, through which light moved, at this time, was thought to be dragged along with the earth through space. Star aberration however disproved this, showing that the aberration was due to lights displacement by the earth's movement and was not dragged along by the atmosphere. French physicist Augustin Fresnel would then develop the partial aether drag theory, suggesting that light traveling through air would not be dragged along, but that as the refraction index of a substance increased, so also would the drag effect increase. Light would therefore travel unimpeded through the air and be uninfluenced by the movement of the earth. Once light enters into glass or some other medium of higher refractive indexes it would therefore be constrained or locked into the new medium and travel at one velocity with that medium (Dennis, pg. 19).

George Biddell Airy later in the 1800's tried to detect the aether using a kind of water filled telescope. The theory was that light traveled slower through water and that while observing star aberration due to earth's movement through space, the aberration angle would change. This also failed; indicating that the aether or medium which light traveled did not exist, adding perplexity to this study (Dennis, pg. 20). In the late 1800's Albert Michelson and Edward Morley went on to disprove the existence of the aether with their interferometer experiment (Dennis, pg. 17).

Willem de Sitter, early in the 1900's discovered that light traveled the same speed in his studies of twin stars; much like Arago's experiments in the 1700's. His discoveries showed the Doppler Effect evident in the shifting of the light spectrum towards the red or blue due to the movement of these stars. Although the speed of light remained the same to the observer, the frequency of the light viewed changed proportional to the speed and direction the star was moving in reference to the observer (Dennis, pg. 21).

These experiments confirmed to scientists of that day that there was no aether. Not only did these tests show that there was no aether, they had also discovered what is now called "Law of the Constancy of the Speed of Light." This measurement of the speed of light is the same no matter what the observer's velocity is what velocity the emitter is, either moving towards or away from each other expressed as "c" and is measured at 299,853 km/sec. or 186,328 miles/sec (Dennis, pg. 30).

Hedrick Lorentz would then try to rectify this strange problem by developing the "Lorentz Contraction" theory (Dennis, pg. 24). This theory showed that the movement of an object causes length contraction of that object in the direction of its movement, therefore causing the speed of light to remain constant no mater what the movement of the observer or object being observed. This is one explanation in how the Muon can reach the earth at high speeds even though its life cycle supposedly would prevent it. The "Lorentz Contraction" due to its velocity at near light speed, changes the space-time environment of the Muon, allowing it to reach the ground.

Albert Einstein would then build on this to develop his "Theory of Relativity" (Dennis, pg. 27). Einstein theorized that movement changed time and space and that the faster an object moves the more time and space around it would change. Gravity also would have a dramatic effect on space and time causing light waves to bend around stars and for time to vary in different gravitational fields. This explains the constancy of the speed of light and the shift in the light spectrum due to gravitational fields and light from objects at different velocities.

Modern day studies seem to verify Einstein's theories beginning with studies on the precession of Mercury (Dennis, pg. 49). Studies of starlight being bent by the suns gravity, gravitational lensing due to the intense gravitational fields of galaxies also validates this theory. Galactic redshift and gravitational redshift is seen in the GPS transmission (Dennis, pg. 47) or radio waves and even the tracking of Venus and Mercury with radar (Dennis, pg. 51). It seems that no matter how you look at it, everything seems to indicate that Einstein was correct.

3

Edwin Hubble's discovery in 1929 showed that while studying galactic red shift that redshift in light from stars and galaxies increase in proportion to their distance from earth. This developed into what is now known as Hubble's Law, expressed by the equation $v = H_0D$. The further we look out into the universe, the faster galaxies and stars accelerate away from each other, indicating that these galaxies and stars were moving out from a single point in space over a long period of time (Dennis, pg. 45).

This discovery lead to the "Big Bang Theory" as scientist began to work back in time to a point where this rapidly expanding universe theoretically began, to a one point in space and time called the "singularity" over 13 billion years ago. In 1964, Arno Penzias and Robert Wilson would discover Cosmic Microwave Background Radiation (CMBR), evidently a relic of the "Big Bang" dispersed evenly across the sky (Dennis, pg. 53).

Studies from observing these stars and galaxies generated new ideas, one being Dark Energy (Dennis, pg. 55). What was the force that acted as a dispersing agent, pushing the galaxies and stars away from each other in deep space; an energy that was repelling against the forces of gravity, forces that should pull everything together in what eventually would be the "Big Crunch"? Another recent observation also suggests the existence of Dark Matter, an invisible mass that would generate enough gravity to keep the rapidly spinning galaxies from flying apart. Existing "known" mass in galaxies such as our own Milky Way, does not have the amount of mass needed to keep stars on its outer edge, moving at great velocities, from flying off into space.

The "Lorentz Transformation" Hypothesis

The failure of Michelson and Morley to detect the aether was puzzling. Soon after the Interferometer experiment, Irishman **George F. Fitzgerald** (1851-1901) professor of experimental philosophy at Trinity College, Dublin, suggested that the negative results of the experiment might have been due to the contraction of the interferometer in the direction of movement. Hendrick Lorentz built on this idea developing the contraction hypothesis based on his Electron Theory stating that any material body is contracted in the direction of its motion. This contraction would then change by a factor relative to the speed of light (Einstein, 1952, pg. 10).

Lorentz's contraction theory would soon serve as the foundation for the mathematics of Einstein's theory of relativity. Einstein developed on this hypothesis making some substantial changes. Einstein modified the Contraction Theory showing that **these changes due to motion do not actually warp the body in motion**, as Fitzgerald and Lorentz had originally suggested, but instead warped or **made changes in space and time around that moving body** (Ditchburn, R.W. pg. 318).

Lorentz's transformations, which he introduced in 1904, formed the basis of Einstein's Special Theory of Relativity. They describe the increase of mass, the shortening of length and the time dilation of a body moving at speeds close to the velocity of light. This may all seem out of the norm as no one or nothing man has yet made even travels close to these speeds. However, this does affect our Earth and us on the Earth. The Earth moving in its orbit around the Sun moves at about 30,000 meters/sec (30 km/sec or 18.5 mile/sec). The contraction of space caused by the high velocity of Earth through space would contract the diameter of Earth by about 6 cm (2.5 inches). This small change may account for Michelson and Morley's negative result by making the source of light and the mirror draw closer together when the system is moving lengthwise nullifying any change due to the earth's rotation and velocity through space.

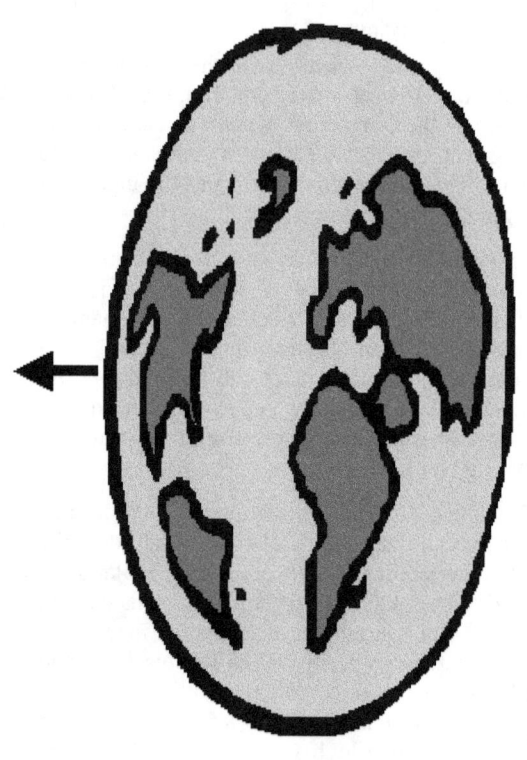

Lorentz Transformation of Time and Space

In three-dimensional space, the Moving System moves with velocity v in the x direction with respect to the Fixed System reference; the moving X' is contracted.

"z" represents height, "y" represents width and "x" represents length of an object. Here, measurements are compared to a fixed system relative to a moving system. Note – Length "x" is not = x'.

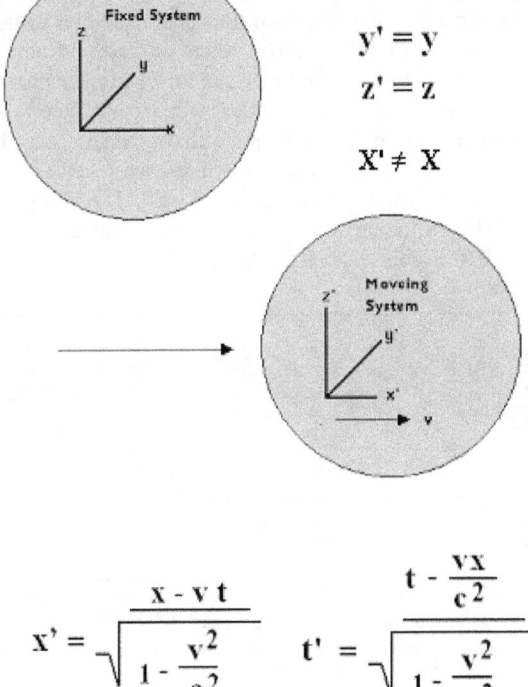

$$y' = y$$
$$z' = z$$

$$X' \neq X$$

$$x' = \sqrt{\frac{x - v\,t}{1 - \dfrac{v^2}{c^2}}} \qquad t' = \sqrt{\frac{t - \dfrac{vx}{c^2}}{1 - \dfrac{v^2}{c^2}}}$$

Lorentz Transformations

Lorentz used an arcsine or inverse sine formula to express these transformations.

$$= \sqrt{1 - a^2/b^2}$$

Inverse Sine Formula

This formula divides the square of two values (such as two velocities) in order to show the ratio between these two values in a range from 0 to 1 (**A**). This ratio is then subtracted from one to return an inverse nonlinear value that cannot be greater than 1 (**B**). This value is then factored with other values that determine the change in length, time or mass variable that changes according to velocity (**C**). The number one or maximum values is then substituted with the value of the speed of light, 186,000 miles per second or 300,000 kilometers per second; the percentage can also be used such as .8 to express 80% light speed.

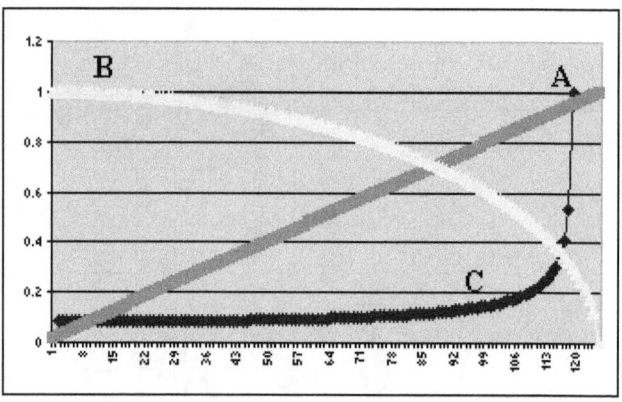

Formula Variations

Stationary System

v = velocity = 0

l = length = 100 ft.

t = time = 10 seconds

c= speed of light

m = mass = 10 Newton's

Moving System 1

If velocity v = .5 X c then

Length of contraction =

.866 x 100 = 86 ft.

Time dilation =

1.15 X t = 11.5 seconds

Relative Mass =

1.15 m = 11.5 Newton's

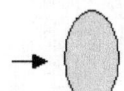

Moving System 2

If velocity v = .999 c then

Length of contraction =

.044 X 1 00= 4.5ft

Time dilation =

22.366 X t = 3.7 minutes

Relative Mass =

22.366 X m =223.7 Newton's

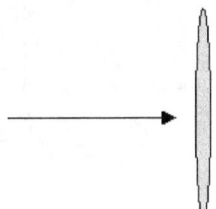

$$l' = l \sqrt{1 - \frac{v^2}{c^2}}$$

Contraction of Space

The "Lorentz Contraction" Theory

Length Compaired with the Velocity of an Object

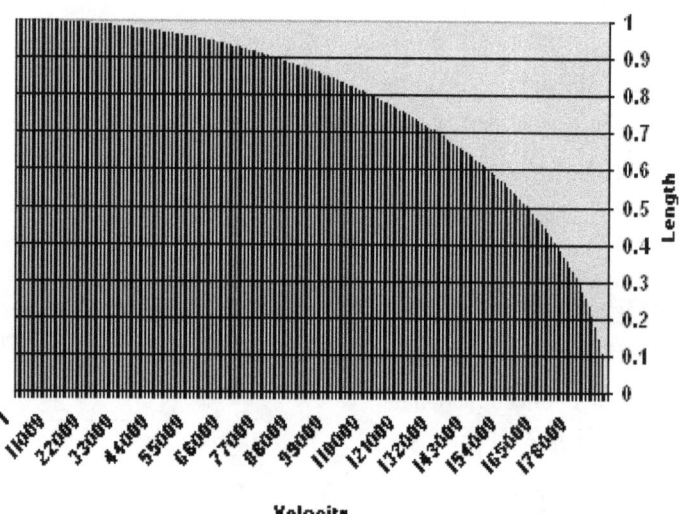

Length Contraction

$$t' = \dfrac{t}{\sqrt{1-\dfrac{v^2}{c^2}}}$$

Dilation of Time

The Lorentz Time Dilation Theory

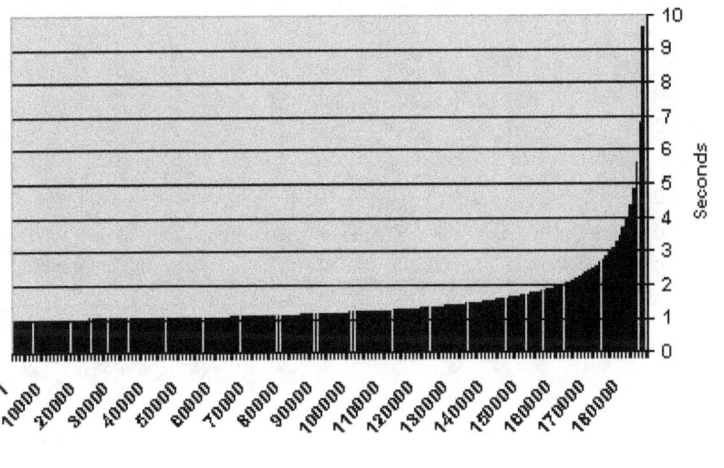

Time Compared with the Velocity of an Object

Time Dilation

$$m = \frac{m_0}{\sqrt{1 - \dfrac{v^2}{c^2}}}$$

Relativistic Mass Formula

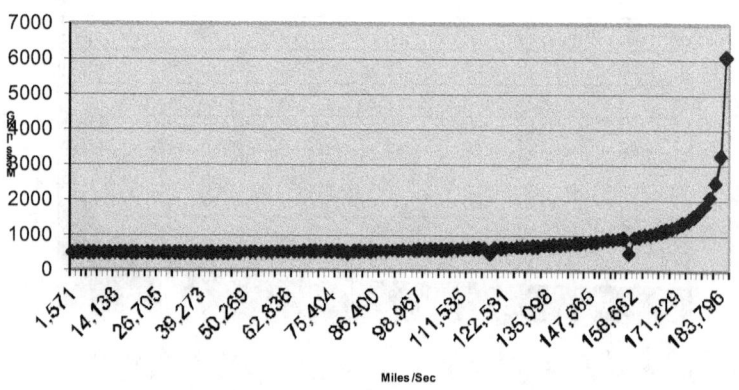

Mass Change in Velocity

Relativistic Mass

Einstein's Remedy

As we saw earlier, Einstein's remedy for this "speed of light dilemma" in the early twentieth century was to alter space and time, building on **Hedrick Lorentz's** Contraction Theory. For example, if a baseball is thrown at 100 mph and is measured by a radar gun that is moving in a vehicle towards the baseball at 50 mph, the reading on the radar gun would read 150 mph. The only way to keep measuring the velocity of the baseball at 100 mph instead of 150 mph is to alter "**space and time.**" We would therefore either have to change the length of the measuring rod, as Einstein would say it, or change the rate of the clock or both. In this case we would have to change the way the radar gun calculates the rate of change in the baseballs position.

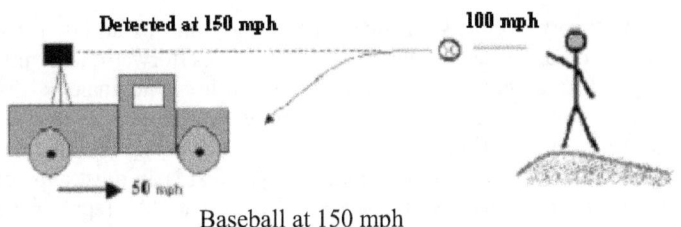

Baseball at 150 mph

This of course does not affect our baseball game; baseball speeds just aren't fast enough. This concept is however very important to understand as we venture further into the future and develop new technologies. If we ever do develop these hyper speeds, what would happen to us at these velocities?

According to Einstein for example, if the measuring rod on the moving object was a yard, it would then have to be perceived to be shorter by a stationary observer (Contraction of Space) and one second to the stationary observer would be less than one second on a clock on the moving object (Dilation of Time). Of course inside that moving system the yard would still be perceived as one yard and one second perceived as one second; no change at all would be perceived in space and time by an observer moving with that system (local space and time). In that way, to a moving observer, no matter what their velocity towards or away from the light source, their readings

13

while detecting the speed of light will always read 300,000 km/sec or 186,000 miles/sec. Shown below is an example of the earth in its orbit around the sun. At certain times of the year, the earth will move at a very high speed towards a specific star, at other times of the year, it moves at a high rate of speed away from that star.

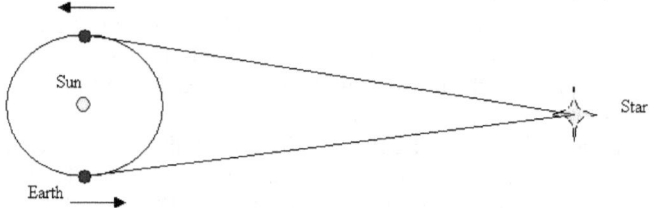

Earth and Star

In theory, velocity changes space and time. The faster the velocity of an observer, the slower he moves forward into time compared to a stationary observer. Einstein used the analogy of a moving train in reference to the stationary embankment beside the train track in his writings, where time and space is slightly different on the moving train because of its velocity. Next is an illustration of Einstein's Co-ordinate Systems A and B. System "A" represents a stationary system and System "B" represents the moving system. The embankment - (System A) in Einstein's Train analogy symbolizes the stationary system while the train represents the moving system (System B) (Isaacson, Walter, pg. 123). The height or vertical plane of three-dimensional space in both systems is represented by y where $y' = y$. The horizontal aspect of width in both systems is represented by z where $z' = z$. The horizontal plane that corresponds to the fore and aft movement in three-dimensional space in both systems is represented by x where $x' = x$ – velocity multiplied by time or x-vt. Here x' is not $= x$ and the length of an object is shorter or compressed in the direction of movement by a specific magnitude the faster an object moves as seem in length x' relative to x, according to the Theory of Relativity.

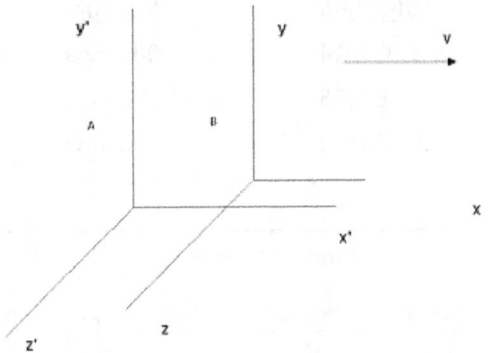

Co-ordinate Systems – Contraction of Space

Illustrated next is the variation in time (Local Time) between three moving systems traveling at three different velocities. System "A" is traveling at 1000 mps (miles per second relative to a stationary observer) where 1 second is equal to 1.000014 seconds. System "B" is traveling at 170,000 mps where 1 second is equal to 2.464495 seconds. System "C" is traveling at 180,000 mps where 1 second is equal to 3. 969143 seconds. Time moves slower in a moving system than a stationary system, therefore theoretically, the faster an object moves the slower a clock ticks. At a very high velocity, say on a spaceship, this time dilation would make it possible for passengers to travel further into the future while aging very slowly compared to people on a stationary object (Born, Max, pg. 317-320). In other words, everyone on earth has aged much more than those in the spaceship

Seconds	Velocity
1.000014	1,000 mps
2.464495	170,000 mps
3.969143	180,000 mps

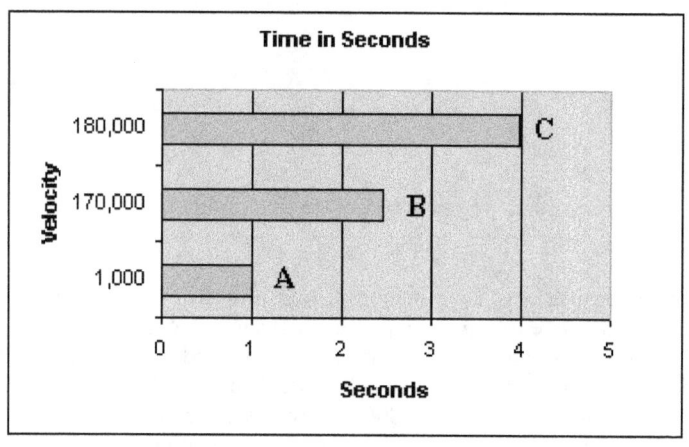

Co-ordinate Systems – Dilation of Time

Blast Off

To test this theory, let us perform what Einstein called, a thought experiment. When? It is the year 2050 and technology has brought us to a turning point. Using a now safe Cold Fusion nuclear power plant and the latest version of the Ionic Thrust Engine (ITE), we are now ready for a venture never before possible. Assembled in orbit around the earth, a five-stage space vehicle is readied for launch with five crewmembers on board. Its fuel is water, accelerated in vapor form to approximately the speed of light by ionizing droplets of the water and accelerating them though a magnetic field. Each stage will "burn" for approximately twenty-four days straight, maintaining an acceleration of 98 ft/sec^2 and a constant force on the crew of 3g's (three times Earths gravitational force) for the entire burn. If you would experience this in a car, this would be an acceleration of approximately 0-66 miles per/hour in one second maintained for twenty-four days straight, a little rough on the human body.

The crewmembers are strapped in and hooked up to devices that provide nutrients and also supply and remove fluids and toxins from their bodies over the next twenty-four days. After blastoff, the crew monitors the spacecraft for a few hours, and then is put into a deep sleep. After these twenty-four days are up, the crew are gently awakened and allowed to adjust to the environment. Then on the twenty-fifth day, engines are shut down and the acceleration stops – the crew are weightless. Crewmembers unbuckle themselves and enjoy this release of pressures on their bodies and are give a day to recuperate.

At the end of their one-day break, one crewmember enters the observation cabin on the first stage booster. The booster is then separated from the space ship; the crew is readied, and then blasts off again leaving Stage One and one observer behind. Like the first stage, this ship also accelerates to 3 g's for twenty-four days straight and the crewmembers again sleep away the time until awakened for the next brief rest. Like previously done, the second stage and one observer are left behind and Stage Three blasts off, accelerating on at 3 g's for twenty-four more days off into space.

This process goes on five times till the fifth stage and its lone observer shut down the engines and cruise weightless in space in a different world. Where are they and how far and fast have they traveled? Let's find out.

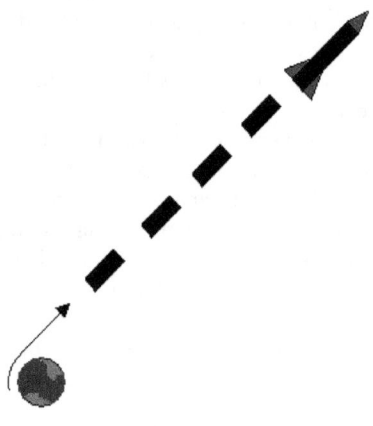

Five Stage Rocket to Light Speed

First burn - How far did the crew travel and how fast are they traveling having maintained 96 ft/sec/sec or 3 g's for twenty-four days straight? Remember, this was 0-66 miles per hour in one second. After one minute or 60 second later, the crew would be accelerated to 3,927 miles per hour and after one hour, 235,636 miles per hour. When twenty-four hours or one day has elapsed, the space ship will be traveling at 5,655,272 miles per hour or 1,570 miles per second. When the first twenty-four days have past and the crew is on their first well needed break, they will be traveling at 37,701 miles per second, one fifth the speed of light.

Second Burn - This next stage blasts off and accelerates for twenty-four more days up to 75,403 miles per second. Third through Fifth Burn - this process continues until the fifth stage completes its burn traveling at a calculated velocity of 188,509 miles per second, faster than the speed of light in reference to the Earth. A total of 121 days have now passed and the spacecraft has traveled more than 60 billion miles.

Now that we have gone this far, it's time to return back to earth. Each stage of this ship adrift in space retained half of its fuel (water) for the return home. The fifth stage and its resident take one day to recuperate and then turn the ship around, firing up the engines for the long return trip back home. Blasting off back towards earth, the fifth stage returns back toward the direction it came from, with a rendezvous of course in store with Stage Four floating aimlessly in the deepest parts of the universe.

Like all of the other separation launches of each stage, Stage Five also accelerates up to and maintains an acceleration of 98 ft/sec^2 or 3 g's. This acceleration however is back in the direction of which Stage Five came from and is in fact a deceleration, facing in the other direction. The effect on the ship and its passenger is however exactly the same, being pinned up against the seat for 24 days straight. What we are doing is actually slowing down Stage Five to Stage Fours velocity and catching back up to it to rejoin the capsule and crew. This will go on for four more link ups as the space ship returns back to earth, taking another 121 days to complete. Each launch decelerates the ship down to the velocity of the prior stage until the ship returns back to earths orbit whole again.

If this is hard to understand, let's put it this way. When Stage Five blasts off and leaves Stage Four behind, it will be traveling after 24 days at 37,701 miles per second away from Stage Four. In space however, this is all relative. Who is to say that Stage Five is moving away from Stage Four? Stage Four could also be viewed as moving away from Stage Five at the same velocity with Stage Five being stationary in space (after it stops accelerating of course). If this were the case, Stage Five would have to use its rocket engines to catch up to Stage Four. This is in fact what we are doing on the return back to earth. With no real reference of what is stationary space, anything not accelerating or decelerating can be considered as stationary. Each stage viewed from another will be traveling at a different velocity – see Relative Velocity chart.

Now, according to calculations, the observer on Stage Five before his return on day 121 was theoretically traveling at 188,509 miles per second relative to the earth. Stage Five would be therefore traveling 2,218 miles per second or almost 8 million miles per hour faster than the speed of light. Can this be?

19

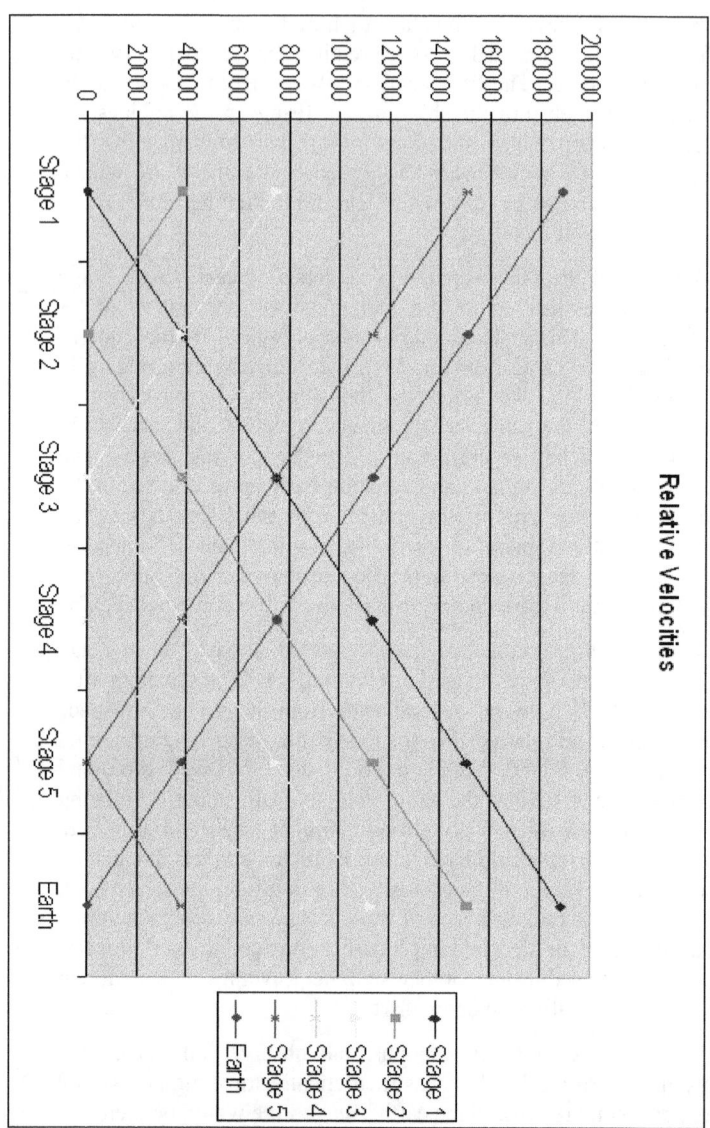

Relative Velocities

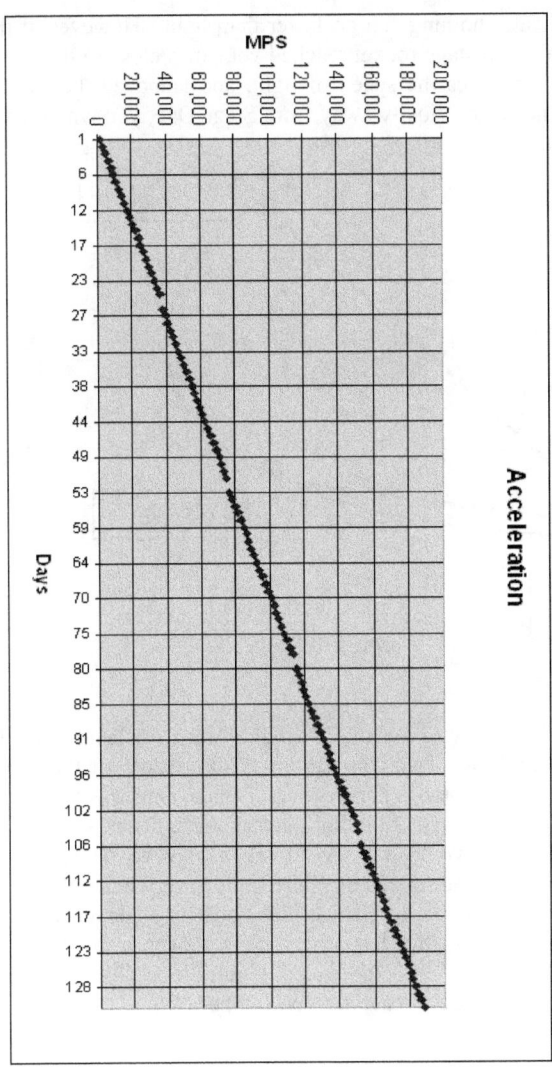

Acceleration

Theoretically nothing can go faster than light. However, if the first stage of a two stage rocket reached velocities close to the speed of light say .95 times the speed of light, and launched the second stage up to the same velocity away from Stage One, you would have this scenario.

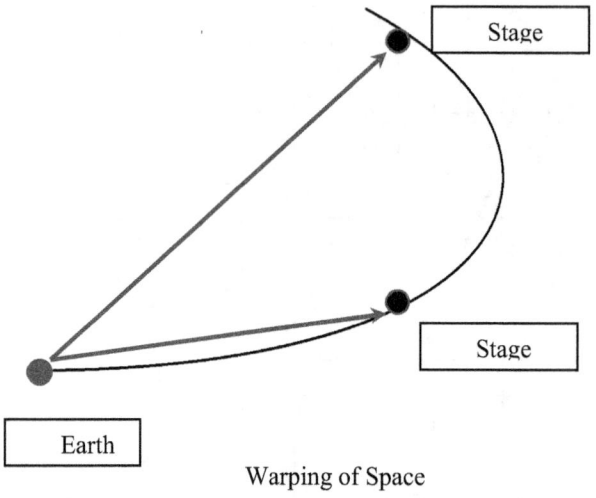

| Stage |

| Stage |

| Earth |

Warping of Space

Einstein's Theory of Relativity would "warp" or bend space in order to prevent Stage Two from going "Warp Speed" or the speed of light. From the Earth's perspective, if Stage One was at .95 the light of speed, Stage Two could not go very much faster, say .98 the speed of light and would have to travel in a curve. The Lorentz Contraction Theory introduced by Einstein does very much the same thing by reducing the rate of change in distance exponentially. This rate of change occurs the faster an object moves away from or towards an observer until the velocity does not change at all, preventing an object from reaching "Warp Speed."

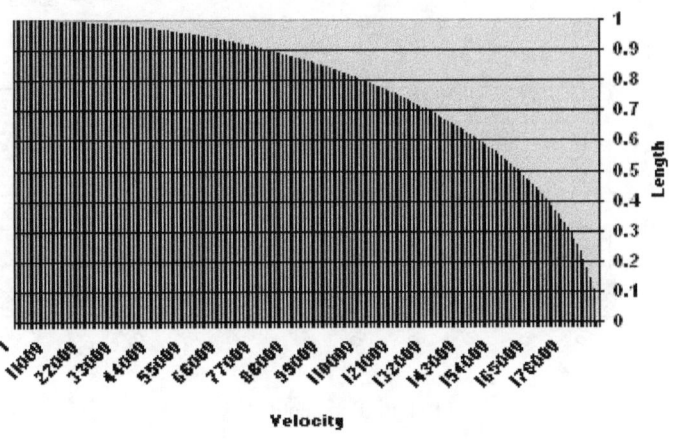

Length Compaired with the Velocity of an Object

Contraction of Space

From the earth's perspective (see chart on following page), with each stage floating in space moving at different velocities relative to the earth, lengths will contract more and more. For example, with a ten-stage spaceship, stage one will be viewed as having very little change. Stage five however will have a contraction of approximately 80% earth's measurements and stage 10 with length contraction around 1% earths measurements. From stage 5's perspective, both the earths and stage 10's lengths will contract to about 80%. Stage 10 will view length measurements of other stages opposite that of earth with earths lengths at 1% and stage 5 at 80% that of stage 10's measurements.

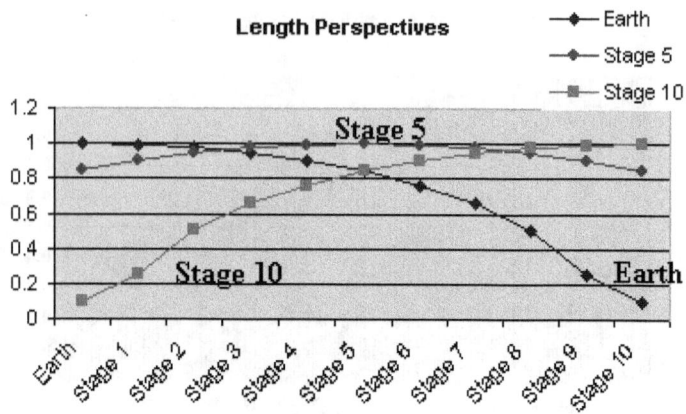

Length Perspectives

In another perspective, space would be compressed in front of a high-speed object similar to a bow wave in front of a boat and be stretched or dilated behind the object. Einstein had suggested that it was space – time that warped, not the moving objects.

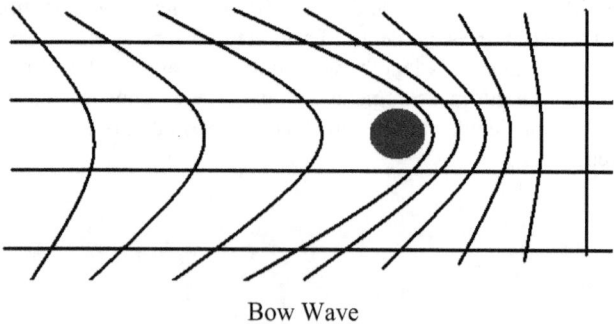

Bow Wave

If time is changed, the distance traveled must also change. Therefore, another perspective is to "warp" time in order to prevent Stage Two from going "Warp Speed." According to Einstein, the Dilation of Time would exponentially effect time the faster an object moves away or towards an observer.

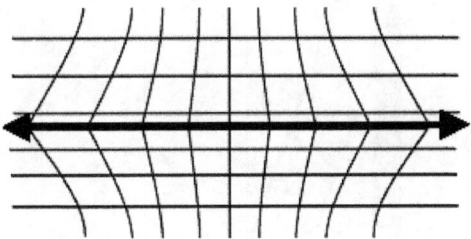

Dilation of Time

Going back to our rocket mind experiment, while viewing this rocketry from earth, Stage 5 hits terminal velocity at day 120, reaching a calculated velocity of 186,291 miles per second relative to the earth. At day 121, if it can continue to accelerate above light speed, it will reach a velocity of 188,509 miles per second, 8 million miles per hour faster than the speed of light. However, while viewing Stage 5 from Stage 1's observer's perspective, Stage 5 can accelerate on to day 146 where relative to Stage 1 it will hit terminal velocity at the speed of light. Remember that from Stage 1's perspective, he is stationary in space and the Earth is moving away from him in the other direction at 37,701 miles per second. The speed of light must have a reference; this reference will always be the observer. If there are two or more observers moving at different velocities in reference to each other and the same light source, the speed of light they detect will always be "c" to each of them.

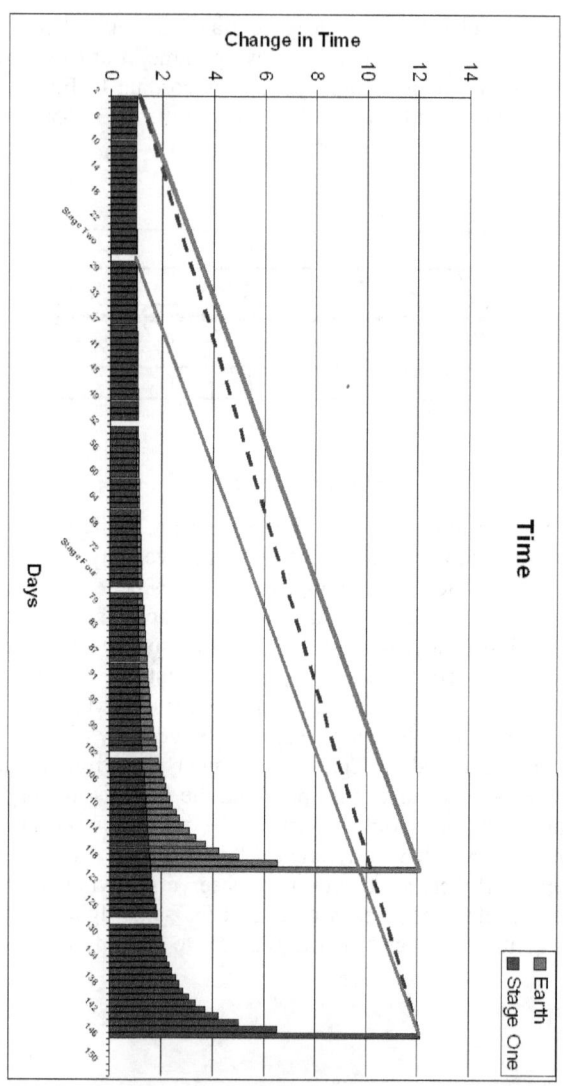

Time

26

In 1904 Henri Poincaré wrote:

"From all these results, if they were to be confirmed, would issue a wholly new mechanics which would be characterized above all by this fact, that there could be no velocity greater than that of light, any more than a temperature below that of absolute zero. For an observer, participating himself in a motion of translation of which he has no suspicion, no apparent velocity could surpass that of light, and this would be a contradiction, unless one recalls the fact that this observer does not use the same sort of timepiece as that used by a stationary observer, but rather a watch giving the "local time"

(Poincaré, 1904, p. 253)

Back to Basics

In the late 1890's, the pace of scientific discovery was colossal. Through the discovery of x-rays and other types of radiation, scientists such as **Henri Becquerel** as well as **Pierre and Marie Curie** began to put together the atomic model. Just after the turn of the century in 1911, **Ernest Rutherford** concluded through his research that a positive charge was confined to the small region at the center of the atom, what he called the nucleus; this was only fifteen years after the discovery of nuclear radiation by Becquerel and the Curie's (Bord, 2000, p. 441). Several years later, in 1913, Danish physicist **Niels Bohr** published an atomic model that better explained the natural atomic spectra using the now well-known "solar system" as a nucleus and electrons in distinct orbits around the atom (Bord, 2000, p. 381). The original model only subscribed to a positive and a negative charged particle model until in 1935, **James Chadwick**, working with **Irene Curie's** work (Marie Curie's daughter) discovered the neutron.

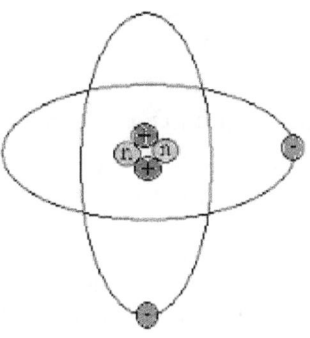

Bohr's Model of the Atom

The electron is the life of the atom. Electrons are believed to travel around the nucleus of the atom at fantastic speeds, some 700 miles per second and higher. Scientifically speaking, the electron is assigned a negative charge and the proton a positive charge, with opposite charges theoretically attracting each other equally. It is this high velocity in orbit that keeps the electron from falling into the nucleus of the atom as a result of the tremendous mutual attraction of the electrons and protons charges.

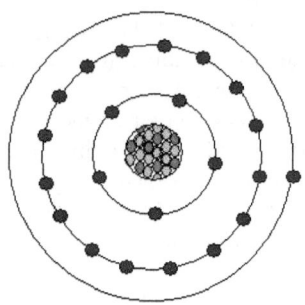

Conductor - One Electron in Outer Shell

Although the forces of the electrons charges are equal to that of the proton, the electron is approximately one trillionth of an inch in size with the proton around 1,800 times the electrons size. The neutron is similar in size to the proton but has no charge. A stable atom of any element normally carries a balance of multiples of these three atomic particles. A positively charged atom will carry more protons than electrons (a deficiency in electrons) and a negatively charged atom will carry more electrons than protons (Veatch, H., 1978, pg. 22-23).

Today, about 155 combinations of these particles are known – called **elements** and their **isotopes**. Elements with a different number of **neutrons** than **protons** may have the same characteristics chemically but are considered an isotope and may be radioactive, behaving differently from a nuclear perspective (Bord, 2000, p. 413).

Some elements such as copper are called conductors and have the ability to share electrons in the outer orbits (shells) of their atoms. As a rule of thumb, elements with one, two or three electrons in their outer orbit are considered conductors, such as gold, silver and aluminum. These electrons move quite easily from atom to atom; this flow of electrons is called electricity. This sharing of electrons also produces the **Covalent Bond** that binds atoms together creating various chemical structures.

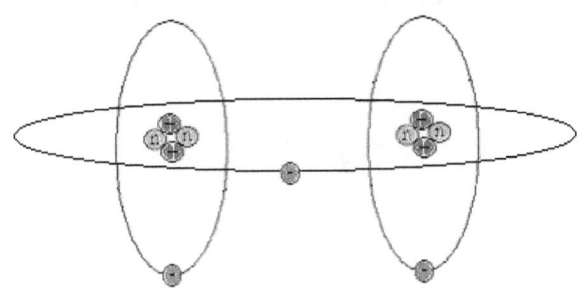

Conductor Sharing an Electron

Other elements have their outer orbits completely full and do not share electrons. These elements are considered insulators and resist electrical currents because their valence shell (outer shell) is full with valence electrons. The more electrons in these outer shells, the more inert the element is. Materials that fall in between insulators and conductors are called semiconductors such as carbon, silicon and germanium and are used primarily in electronic circuitry and integrated circuit chips (IC) used in electronics today (Evans, Alvis, pg. 2).

Whether in chemistry or electronics, the electron is a source of energy that can be tapped. Electricity can be used to produce heat, magnetic energy in motors and electrochemical reactions such as chrome plating. Electricity is also used to create light and various types of electromagnetic propagation such as radio waves, microwaves, x-rays and gamma radiation. It is in understanding the electron that we can start to understand electromagnetic phenomena, energy and the fundamentals of light.

Energy

What is energy? Many eastern religions and even the new age movement speak of pure energy. Is there such a thing as "Pure Energy"?

Scientifically speaking, energy is a variable quantity that is a characteristic of an object. Energy is the ability to do work or that which is transferred when work is being done (Bord, 2000, p. G-1). There are several ways to describe energy; kinetic, potential, gravitational, electromagnetic, chemical and nuclear. Energy can be transformed from one form of energy to another such as light into heat, but cannot be created nor destroyed. This is known as the **Law of the Conservation of Energy** written in **Sir Isaac Newton's** "***Philosophiae Naturalis Principia Mathematica***," Latin for "Mathematical Principles of Natural Philosophy"- one of the most influential scientific works ever published.

Newton's Law of Conservation of Energy

"Energy cannot be created or destroyed, only converted from one form to another. The total energy in an isolated system is constant"

(Bord, 2000, p. 98)

Let's start with some energy formulas:

Velocity = Distance/Time; The rate of change in the position of an object over time.

Acceleration = Velocity/Time or Distance/Time^2

An example would be meters/second^2. Change in velocity of an object over time, over time.

Momentum = Mass times Velocity or P = mv; the mass of an object multiplied by its velocity.

Inertia = A resistance of an object to the change in its state of motion. F = ma or Force = mass times acceleration

As we can see, all of these formulas contain an object. Objects have mass, therefore "energy" simply put is a quantity associated with the movement of mass either kinetically (active) or potentially.

As a ball is thrown, it is accelerated by the person throwing it to a specific velocity and has a certain momentum; the ball then decelerates and falls to the earth. The energy put into the ball dissipates through friction in the air and gravity. The remaining energy the ball has then is transmitted into and dissipated through its contact with the ground as vibrations and as sound through the air.

A rocket shot into space will remain at the velocity it was accelerated to once the engines are shut down as we saw in Blast Off. It will stay at this velocity and direction of travel until acted upon by another force such as gravity or friction due to reentry into the atmosphere.

Newton's First Law of Motion

"Every object in a state of uniform motion tends to remain in that state of motion unless an external force is applied to it"

(Bord, 2000, p.47)

Einstein's Most Famous Equation

As we look at Einstein's most famous equation we also see that mass has to be part of the energy equation.

$$E = MC^2$$

Here E is energy, M is mass and C is the speed of light at approximately 300,000 kilometers per second or 186,000 miles/second - identical to momentum.

Mass x Velocity or $P = mv$ (Born, Max pg. 283-286).

34

Subatomic Particles

There is a difference however in how subatomic or elementary particles react with other particles and how they dissipate their energy. Unlike our world, where energy such as sound is dissipated through many different *medium*s and distributed into billions of particles, energy from subatomic particles can only be distributed and dispersed into fixed packets or bundles of energy that cannot be divided (Polkinghorne J. C., 2000, p. 6).

In the twentieth century, studies revealed that light could cause atoms to emit electrons. In 1900, while studying **Black Body Radiation** (BBR), **Max Planck** (1853-1947) found that atoms emitted energy in multiples of a certain energy unit, E or 2, 3, 4… and so on, multiples of E; it was not possible to emit a fraction of a unit (Polkinghorne J. C., 2000, p. 5). Einstein suggested that light energy may not be evenly spread across a given area but concentrated in small "packets" or "quanta". The energy transferred from one particle to another had to be completely transferred, all or nothing.

This localized energy was coined the "photon" (Kafatos, M. 1990, Pg 29). The "aether theory" was thrown out and the "corpuscle theory" also known as the "particle theory of light," originally suggested by Newton, was then reintroduced. The "aether theory" had too many questions that could not be answered; the "particle theory" solved many of these questions and then became the preferred theory of the scientific community.

Subatomic Expressions of Energy

Subatomic particles such as electrons demonstrate the energy they obtain in several ways.

Spin – Where the particle has angular momentum, a conserved quantity based on its mass and speed that will continue unless acted upon by another force (Polkinghorne J. C., 2000, p. 96). The spins of all known particles consist of a half spin or whole spin or a multiple of that factor, for example 0, ½, 1, 3/2 … etc. This energy is transferred from one particle to another in fixed quantities in particle-like packets of energy called quanta (Bord, 2000, p. 458).

Velocity – Where a particle has orbital angular momentum in orbit around an atom. This energy is also transferred from one particle to another in fixed quantities in particle-like packets of energy called quanta; as an electron absorbs a "quanta" of energy, its velocity and energy level change causing it to "jump" from a low orbit to a higher orbit. Likewise, as an electron emits or looses a "quanta" of energy, its velocity and energy level again changes causing it to "jump" from a high orbit to a lower orbit (Bord, 2000, p. 381).

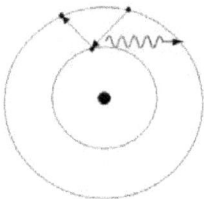

Electron Releasing a Packet of Energy

Frequency – The rate at which a particle oscillates per second (Polkinghorne J. C., 2000, p. 213). This frequency is proportional to the wavelength of the energy it disperses. This wavelength is the measured distance of a cycle from beginning to end. This electromagnetic energy wave travels at the speed of light (29,979,245,800 centimeters per second) and their frequency and wavelength can be determined by these formulas:

Frequency = $\dfrac{\text{Speed of light}}{\text{Wavelength}}$

Wavelength = $\dfrac{\text{Speed of light}}{\text{Frequency}}$

Speed of light = λ x Frequency   Wavelength = λ

The amount of energy a particle such as a photon has depends on the frequency at which it resonates. $E = hf$ where E is energy, h is Planck's constant at 4.136×10^{-15} Electron Volts (eV) and f is the frequency that the particle oscillates at in Hertz (Bord, 2000, p.375).

Examples:

Red light: f = 4.3 X 10^{14} HZ (Hertz) at **1.78 eV**

Blue light: f = 6.3 X 10^{14} HZ at **2.61 eV**

X –ray: f = 5 X 10^{18} HZ at **20,700 eV**

An electron in orbit is balanced between electrical attractive forces that pull it towards the nucleus while its centripetal acceleration caused by angular momentum keeps the electron from falling towards the nucleus. In the lowest orbit, orbit one; an electron has the smallest amount of energy possible. As it gains energy the electron moves out to larger orbits. Each "jump" to another orbit is caused by a gain or loss of packets of energy called "quanta" – Planck's constant. Each jump called a "radiative transition" causes the electron to radiate or absorb energy and change orbit. The larger the transition or "jump", the higher the frequency of the energy radiated. For example, the transition from orbit 3 to orbit 2 may give off the lower frequency red light burst whereas a "jump" from orbit 6 to 2 may give off a high frequency violet light burst. Electromagnetic spectroscopy analysis of these light burst show patterns that are unique to specific elements and can be used as fingerprints in determining the element present by the light they radiate.

The speed of an electron, as calculated by French physicist **Louis de Broglie** (1892-1987), in the smallest orbit of a hydrogen atom is about 2.19*10^6 m/s or meters per second (Asimov, Isaac, pg. 86). The speed of light is 2.998*10^8 m/s or 299,800 kilometers/sec (186,000 miles/sec). The mass of an electron is approximately 9.11*10^{-31} Kg. Therefore the momentum of an electron in the smallest orbit at 2.19*10^6 m/s is 1.995*10^{-24} kilogram-meters per second (Bord, pg. 385).

If this momentum is transitioned to the speed of light though the photon, the mass of the photon would be the energy or momentum of the electron divided by the velocity of the photon which is the speed of light or (1.995*10^{-24})/(2.9979*10^8) = 6.655*10^{-33} Kg (compared to the electrons 9.11*10^{-31} Kg). This seems to suggest that the photon would be 136.89 times smaller in mass than an electron and 251,532 smaller than the mass of a proton. Likewise, if a photon's mass is 6.65496 *10^{-33} Kg. and its velocity is the speed of light at 2.998*10^8 m/s then = 1.995*10^{-24} kilogram-meters per second – enough momentum to cause an electron to "jump" from one shell to another in an orbit around an atom or the affect of "h", a Planck constant or "quanta".

The equation $E = MC^2$ indicates that the energy of a photon depends on its mass multiplied by its velocity (the speed of light) or E (photon) = mv^2. In reference to the speed of light this would be E = c (mv) or E = cP where (mv) or P is momentum.

There is a problem though; according to Einstein, as a particle increases in velocity closer to the speed of light, its mass increases exponentially – called relativistic mass compared to mass at rest. For example if an object whose mass is equal to 500 kg approaches the speed of light, say 90% the speed of light, its mass would increase to approximately 1143.8 kg - more than double its mass.

$$m = \frac{m_0}{\sqrt{1 - \frac{v^2}{c^2}}}$$

Relativistic Mass Formula

According to physicist, Satyendra Nath Bose, the photon must be different from other subatomic particles in that it must have zero mass at rest in order to go the speed of light (Isaacson, Walter pg. 328). As the velocity (here referenced at c, the speed of light) of a particle decreases, the mass of the particle also must decrease to the point where it has no mass at rest (Ditchburn, R.W., pg. 343).

All these fifty years of conscious brooding have brought me no nearer to the answer to the question, 'What are light quanta?'

Albert Einstein   (Isaacson, Walter, pg.101)

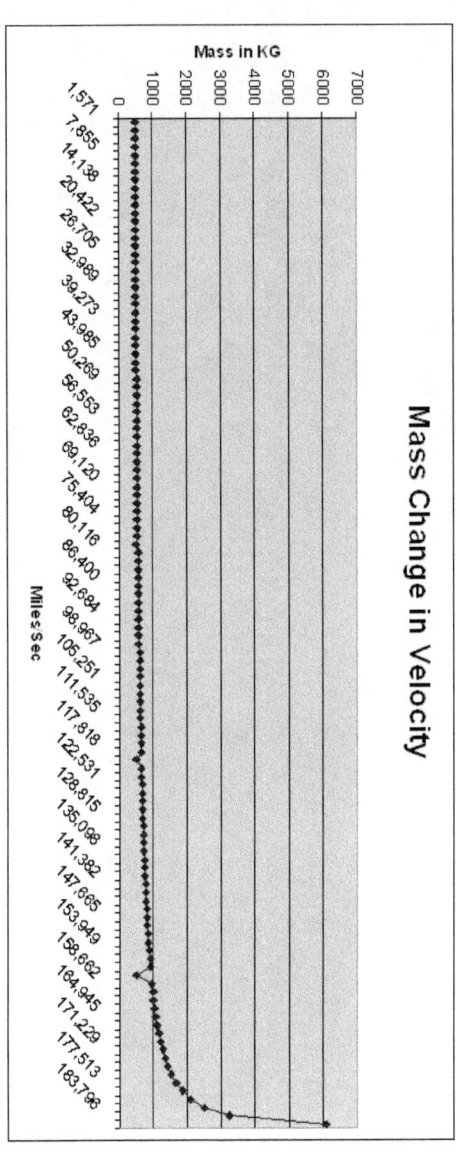

Mass Change in Velocity

39

Questions
and
Theories

Quantum Questions

The following examples will not use the Lorentz Transformations whose numbers exponentially compound, but rather standard multiplication and division for simplicity sake, however the principle is the same.

Let's go back to the first illustration in Einstein's Remedy with the Earth moving towards and away from a star earlier in this book.

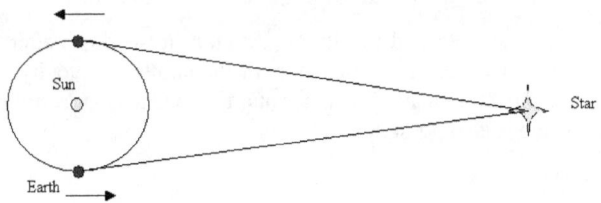

Movement of the Earth

Earth towards Star:

186,000 m/sec +18 m/sec = 186,018 m/sec

Earth away from Star:

186,000 m/sec - 18 m/sec = 185,982 m/sec

Our dilemma here then is that:

   186,000 m/sec +18 m/sec = 186,000 m/sec

And

   186,000 m/sec - 18 m/sec = 186,000 m/sec

One approach to change this is to change time:

Lengthening the second or Time Dilation

1 second = 1.00009674 <u>seconds</u>

Where 186,018/ 1.00009674 = 186,000m/sec

Another approach to change this is to change space:

Shortening the mile or Length Contraction

1 mile = .999032 <u>miles</u>

Where <u>186,018</u> miles X .999032 = 186,000 miles

Note that both end results are 186,000 m/sec no matter what increase or decrease in the detectors or the emitter's velocity. Also note that any combination of space and time values that result in the same value is also possible.

Analysis:

Now let's view this with a spaceship. As the spaceship moves towards the star at a high rate of speed, in order to compensate for the speed of light coming from that star, time and space must change. The spaceship at this velocity **contracts** making everything on spaceship in the plane moving towards the star shorter. Time also is **dilated** allowing for the speed of light from the star to remain at "c". Also from the time perspective, to the stationary observer in space, a clock on the spaceship must then tick slower.

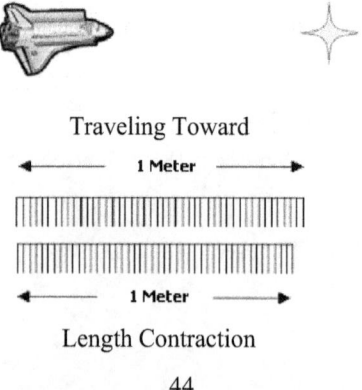

Traveling Toward

◄────── 1 Meter ──────►

◄────── 1 Meter ──────►

Length Contraction

Now, let's move in the opposite direction. As the spaceship moves away from the star, the spaceship theoretically must stretch in the plane moving away from the star to compensate for the speed of light. Time must also then be must compressed to compensate for the constant speed of light. From the time perspective, to the stationary observer in space, a clock on the spaceship must then tick faster.

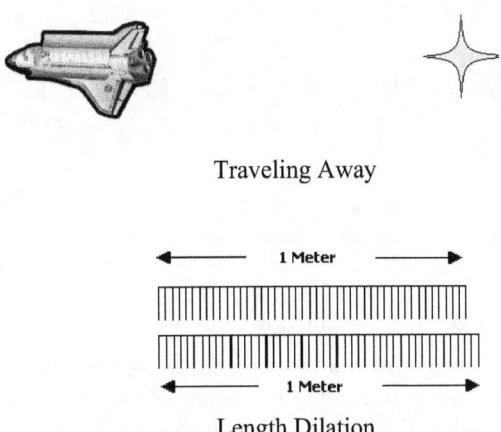

Traveling Away

Length Dilation

In another thought experiment, let's put two spaceships with light measuring devices onboard moving in opposite directions relative to a light source (star). These two devices will travel at a high rate of speed in opposite directions reading the stars light. We will observe these two devices from a stationary observer's viewpoint.

Observing Two Moving Objects

Device No.1 moving towards the star and device No.2 moving away from the star are both reading the same light from the star. Device No. 1's length will contract to compensate for the speed of light and its clock ticks slower than the stationary observer. Device No. 2's length will have to be stretched and its clock will have to tick faster in order to compensate for the speed of light in reference to the stationary observer. These two devices will in fact be emulating the Earths motions towards and away from a star (at different times of the year) at the same time as in our first example.

Now, let's look at it from a different perspective. We will now only have one moving measuring device observing the light from two stars.

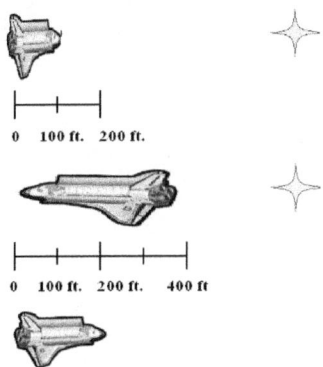

STAR 1

photons

moving device

stationary observer

STAR 2

photons

Observer 2

To the stationary observer, the clock on the device moving towards star No.1 ticks slower and its length contracted to compensate for the speed of light emitted from star No.1. However, the stationary observer also must perceive the clock on the measuring device moving away from star No.2 as ticking faster and its length stretching to compensate for the speed of light from that star. How can this be? From the same observer, the same object is both speeding up and slowing down and at the same time contracting and stretching to compensate for the speed of light. This appears to be a paradox. We must remember that according to Einstein, it is the space and time around an object that changes, not the object itself.

0   100 ft.   200 ft.

0   100 ft.   200 ft.   400 ft

Length Contraction and Dilation

47

A Challenge to Einstein

It was the Michelson Morley experiment with the interferometer that first proved that the aether did not exist. At first glance one might want to challenge the Michelson Morley Experiment by pointing out that the experiment would not work. It would not work simply for the fact that one beam crosses the aether wind at 90^0 and the other beam flows directly into the aether wind. It is being slowed down at first, but after being reflected off the mirror is sped up while flowing with the aether for a null effect. Remember that this experiment is testing a closed loop and that the end result is the average of the round trip speed of light over a short distance.

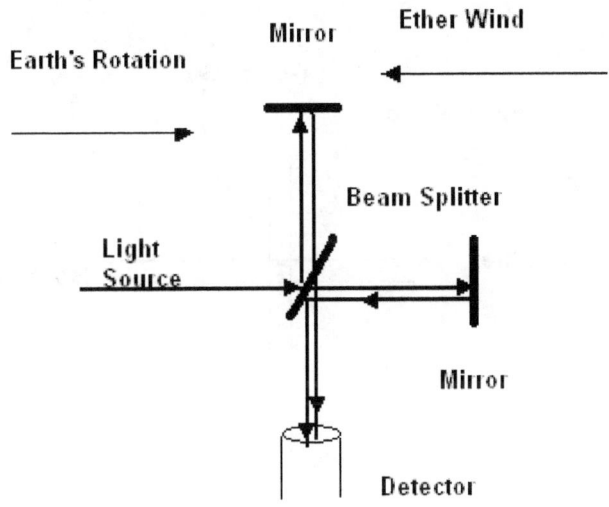

The Michelson Morley Experiment

Let us use the flights of some aircraft for an example to emulate this experiment and compare two flight plans, both 2,500 miles one way. For the first aircraft, let's assume that a flight from Miami, Florida to Goose Bay, Newfoundland, Canada is both 2,500 miles one way and directly north in heading. For the second aircraft, the flight is also 2,500 miles from Boston, Maine to Seattle, Washington and is directly west in heading. The wind is straight out of the west at 100 knots at altitude. Both aircraft will maintain 500 knots airspeed the whole flight at the same altitude.

Using our aircraft navigation computers we take off with the two aircraft at the same time and fly round trip from Miami to Goose Bay and back and from Boston to Seattle and back, at the same air speed and wind conditions, timing both aircrafts round trip flights.

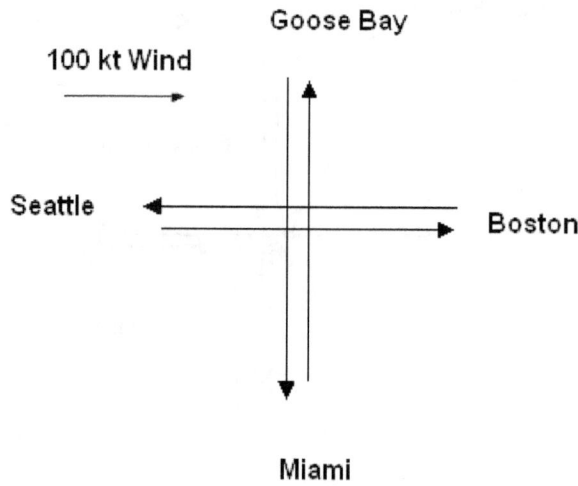

Flight Path

The aircraft flying to Goose Bay averages 510 knots ground speed going to Goose Bay and 510 knots ground speed back to Miami with a 90^0 crosswind both ways. It took 9.804 hours round trip to complete the flight. The second aircraft had to first fly into the wind and then return with the wind. Its first leg to Seattle took 6.25 hours while it fought the wind; the second flight back to Boston, however, only took 4.166 hours flying with the wind. The difference might surprise you; round trip for this second flight was 10.4166 hours, 37 minutes longer than the north south flight. Why the difference? Both flights had to fight the same wind and both flights were 5,000 miles round trip at the same airspeed! The reason? Diminishing returns! The harder the wind, the more time lost. At one point as you raise the bar, the wind will be at a higher velocity than the aircraft can fly and the aircraft will in fact fly backwards in reference to ground speed.

How does this relate to the Michelson Morley Experiment? The Michelson Morley Experiment was an experiment whose intent was to discover the "aether wind." As we can see in our aircraft experiment, there was a difference, but what if the wind was only one or two knots instead of one hundred knots? Would we have noticed the difference? How fast is the "aether wind"? Would the wind be one mile per hour? Would it be five thousand miles per hour? If this experiment would measure anything, how about pointing it up and down? Would it measure the effect of gravity on the light in the vertical plane? Isn't gravity, after all the "aether wind" anyway? Recent studies have used high precision atom interferometers to measure local acceleration due to gravity.

The "Big Bang"

Hubble's discovery in 1929 of an expanding universe lead scientist to try to determine what happed in the past by looking at the universe backwards in time, some 10 to 20 billion years ago to a space-time "singularity" where everything was "one". All matter would come from a certain point where measurements were infinite, as if in a black hole (Asimov, Isaac, pg. 305).

This is how it may have started:

In the beginning, there may have been no such thing as time and space; then something happened. An immense release of energy occurred; a critical stage was reached, possibly similar to "critical mass" in a nuclear event out of the "singularity". A point was achieved in which the accumulation of mass and thermonuclear energy could no longer be sustained and a massive sub atomic nuclear like explosion occurred. In this huge explosion, existing mass is blasted into the smallest of particles as hot plasma like gas out into the far reaches of space. Huge amounts of energy are released as these smallest of particles ricochet off each other at enormous velocities, dissipating their energies out into the blackness of the universe.

Over time this energy continues to spread out from the center of the energy burst. As this outward energy dissipates into the regions beyond and the gases begin to cool, a reversal of the flow or an "implosion" of these particles of mass begins to take place, back toward the center of the blast occurs. As time passes and as the particles energies and spins start to subside, higher energies and pressures now exist to the exterior of this spherical blast, pushing these small bits of matter together forming a denser pocket of matter in the low pressure center. Now cooler at the center, where the blast first took place, these "particles" now begin to go through a **phase transition**, reaching a *critical point* and precipitate or condense, forming a denser core – the beginnings of subatomic particles and atoms.

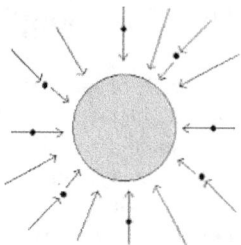

Condensation of Particles

One analogy that can be used is to look at these "particles" as low-pressure systems with pressure gradients such as isobaric lines that slowly decrease in pressure from the outside in. Just as miniature hurricanes or "black holes", material or energy in this material slowly spirals in towards the center of this blob of spinning mass.

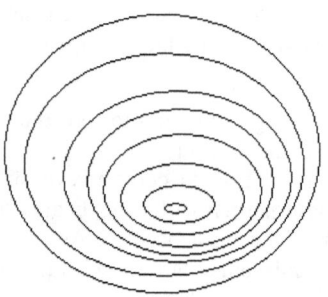

Pressure Gradients

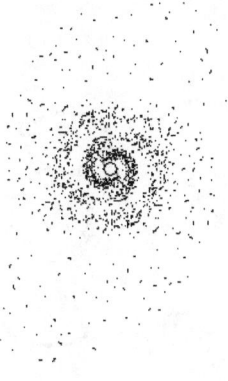

Low Pressure System

Over time, as energy disperses out into space, more and more of these "low pressure systems" begin to form. Elementary or fundamental particles flowing into these systems are spread out over great distances, spiraling in towards the center of these condensing storms. As these different systems come closer together they begin to react to each other even at great distances, as the space between them is not empty. Low-pressure systems spinning in the same direction repel each other even though far apart, bouncing off each other like tops spinning the same way. Systems spinning in opposite directions move together like gears in a clock, attracting one another and move in unison.

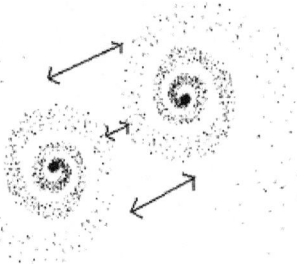

Like Spins Repel

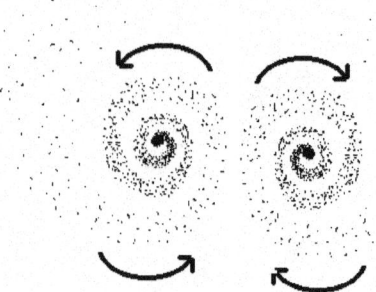

Opposite Spins Attract

    As these systems or particles begin to interact with one another they transfer energy, bouncing off of, spinning with and spinning around each other until equilibrium of energy is attained. Various sizes of particles of different spins pair up with particles of opposite spins creating balance. Smaller particles of less mass with spins opposite of larger particles attract each other causing the smaller particles to begin to spin around the large particle, entering an orbit. The orbiting particles become the life of these systems, speeding up and slowing down as they absorb or give off energy. The same number of particles in orbit balances the number of particles in the nucleus being orbited and an atom is formed.

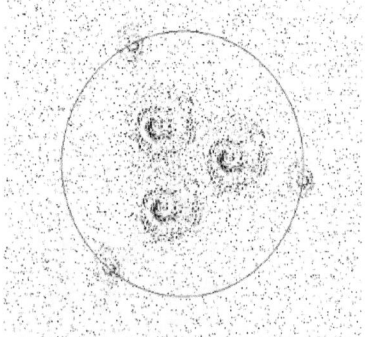

Particles Moving in Unison

The amount of energy given off or absorbed is directly proportional to the size of the particle and its angular momentum in orbit around the central system or nucleus. This amount of energy is a fixed value, a packet of energy, based on the velocity of the orbiting particle in the lowest possible orbit. This energy however, can be passed on to other particles at multiples of that value. As these orbiting particles absorb energy, they increase in momentum and move further away in their orbit around the nucleus. Also as these particles give off energy, they decrease in orbital angular momentum and move closer to the nucleus.

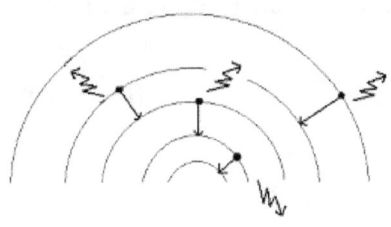

Release of Energy

As energy emitted from these systems is dispersed in these whole value "packets" of energy, they are likewise absorbed as whole packets of energy in other systems, even at some distance. The constant giving off and absorption of this energy makes the particles in orbit move in and out in their orbits causing the nucleus to oscillate forming a wave pattern or "matter wave". This constant waving in and out of the orbiting particles disturbs the cloud of tiny particles that fills the space between these particles causing other orbiting particles even at great distances to resonate in unison. Particles in different orbits resonate at different wavelengths dissipating energy at various frequencies through the space around them.

Simply stated, the hypothesis here is that existing matter is made up of the smallest of particles that are gradually condensing as their energy dissipates out into the universe. These particles become the atoms, protons, neutrons and electrons that create all things in our universe. Energy is simply the movement of these particles in various ways and the interaction of these particles with each other even at great distances. The space between subatomic particles such as electrons and protons is not empty and the movement of one particle will ultimately move other particles transferring energy from one area to another, similar to waves in the sea, at the speed of light even though far apart.

According to the general theory of relativity, space without aether is unthinkable; for in such space there not only would be no propagation of light but also no possibility of existence for standards of space and time.

Albert Einstein (May 1920)

(Isaacson, Walter pg. 318)

Particles

As we saw previously, particles as we know of as electrons, protons and neutrons are possibly condensates of elementary or fundamental particles. These condensates may be formed by pressure gradients created by the cooling of a gaseous mass made up of these particles. During cooling, these condensates reach a saturation point and transition phase, bonding together at a specific energy level into a unified particle. Similar to the dew point, this may occur when specific pressures and temperatures are reached and "droplets" form. These droplets can only exist as a quantity of these elementary particles at specific energy states. These particles can be in a stable state as condensates of cooled fundamental particles formed into protons and electrons. They also may be manifested as moving corpuscle particles, for example, a photon that is produced by pressure waves created through compression by energy moving in the medium.

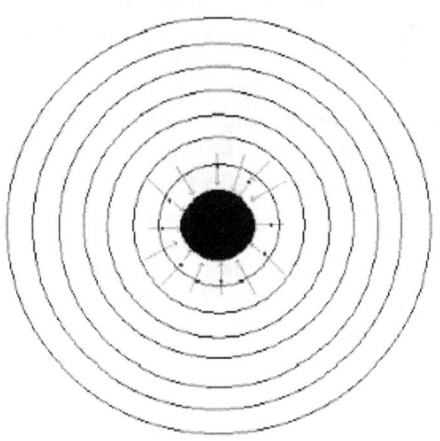

Condensates of Elementary Particles

These corpuscles or discrete particles are formed by pressure waves from energy traveling through an elastic *medium* of elementary particles. Again, a wave is defined as *a traveling disturbance of coordinated vibrations that transmit energy through a substance without net movement of that substance.* According to Webster, *A vibration propagated from **particle to particle** through a body or **elastic medium**, as in the transmission of sound.* These discrete particles are created by energy passing through these relatively stationary particles.

These corpuscles are created by changes in the density of this *medium* from waves of particle motion in a specific direction. Since these discrete particles can only exist under specific energy conditions, they remain confined to only regions of pressure or pressure gradients that create this phase change. In the form of an electron, this energy of moving waves of particles (spherical standing waves possibly) will be confined to a specific pressure level in the *medium* that exists as rings or shells around the nucleus of an atom. As energy is absorbed or emitted from these pressure gradients, existing electrons will dissipate into the *medium* of the region they are presently in and reform as a condensate of energy waves in particles of a different pressure gradient or shell surrounding the atom – disappearing and reappearing like dew drops. Electrons may also be viewed as a sort of hybrid, a stable condensate particle yet wave-like in form. Energy given off or absorbed by these electron waves is only transferred in fixed "packets of energy", allowing these energies to only exist in specific shells specified by their momentum; the energy of the smallest shell or orbit possible, contains the energy of one of these "packets of energy". Each shell would then be created by multiples as if harmonics of this energy.

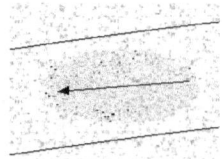

Electron as a Corpuscle of Energy

The energy that is being emitted or absorbed by these regions surrounding the atom is in the form of another discrete particle called the photon. This particle also exists as a pressure wave through the *medium* that can only exist under specific conditions as a "packet of energy". This packet of energy is given off by an electron as a pulse of energy in a given direction. The electron or wave of energy caught in the orbit of an atom that is emitting this energy then dissipates and reemerges as a wave of energy in a lower shell closer to the atom. The wave of energy given off from the electron as a pulse, then moves as a corpuscle of changing density in the *medium* sea of elementary particles in a straight line at the speed of light. This energy continues in that direction until it is absorbed by another electron in the orbit of another atom far away. Since energy cannot be created or destroyed and this particle can only exist under specific energy levels, energy is concentrated in one direction and cannot be dispersed as if moving through a tunnel. No "elementary particle" of this *medium* continues to move in this direction; they simply oscillate in place passing energy from one particle to another, compressing together into a denser mass in a chain reaction in one direction at the speed of light.

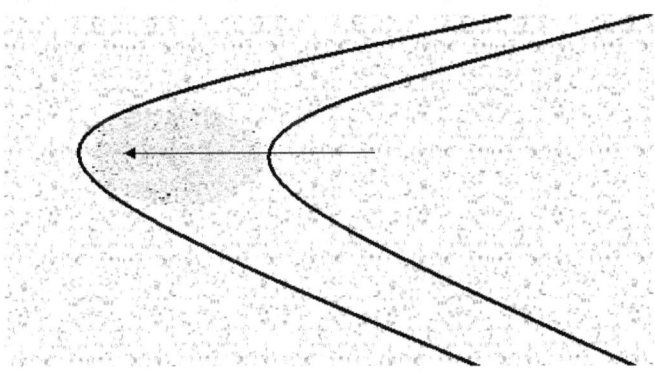

Photon as a Corpuscle of Energy

The photon as a particle, unlike the electron and proton, can only exist as a wave and has no mass of its own, just as a wave in the ocean cannot exist by itself. The photon and the wave on the ocean are simply the manifestation of energy – the movement of mass. The mass of the photon in a snapshot of time is the sum of "elementary particles of the medium" packed together in the wave front and may vary according to the density of the medium and the velocity of the wave through the medium. Its momentum or energy will vary as summations of "packets of energy". This process is continuously generated as electron waves surrounding atoms constantly absorb and emit this energy causing a back and forth waving of these particles in the *medium* of elementary particles. The different intervals or timing of these cycles depends on the energies and the type of atoms giving off this energy, producing different wavelengths of photon waves or colors of light detected in our eyes.

Electrons and photons do not only travel at high velocities in a given direction or orbit; they also oscillate back and forth at a given rate as they move in this direction. The energy of these electrons and photons can be mathematically represented in wavelengths. The lowest orbit possible would hold one packet of energy known as the Planck Constant and one full oscillation of the electron in its orbit would be explained as one full wavelength. An electron in higher orbits will have a higher angular momentum, therefore a higher frequency of its oscillations and a shorter wavelength. The formula $p = h/\lambda$ can be used to show the momentum or energy of electrons in each orbit. Here p is the momentum, h is the multiple of Planck's constant in whole numbers and λ, the Greek letter lambda is the wavelength of the electron (Kafatos, Menas, pg. 33). The formula expressed as $\lambda = h/p$ shows the wavelength of the electron based on the multiple of the constant divided by the momentum of the electron (mass multiplied by velocity, mv). Energy released by the electron changes its momentum, thus its wavelength. This energy is transferred in another waveform out of the atom in the appearance of a photon of a specific wavelength and energy level.

Momentum or "p" of an electron in the <u>lowest possible orbit</u> is equal to the mass of an electron of $9.11*10^{-31}$ Kg multiplied by the velocity of the electron in this orbit of $2.19*10^{6}$ m/s (meters per second) which is calculated to be about $1.995*10^{-24}$ kilogram-meters per second. The wavelength or λ, can now be calculated with the energy of one Planck constant (h) to be 6.63×10^{-34} J-s (joule seconds). Since $\lambda = h/p$, then:

$\lambda = 6.63 \times 10^{-34}$ J-s/$1.995*10^{-24}$ kilogram-meters per sec.

or

$\lambda = 3.32 \times 10^{-10}$ meters or .332 nm (nanometers) the de Broglie wavelength discovered by **Louis de Broglie** in 1923 also know as the Matter Wave (Bord, pg. 385).

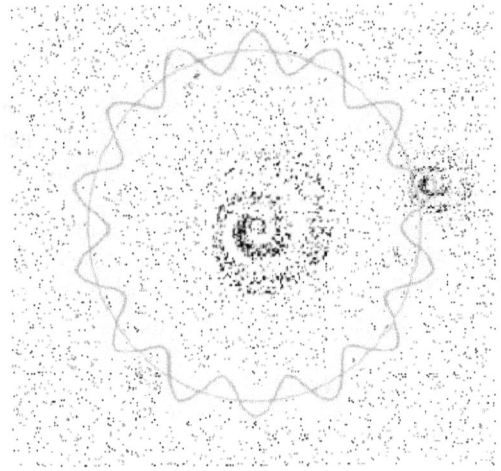

The Matter Wave

Particles whether they be waves of particles or particles as condensates are still created out of this medium and this medium is the foundation of mass and is mass. As particles approach each other, pressure gradients between the particles decrease, causing the differential in pressures around these particles, creating an invisible field that draws them together. Therefore photons and electrons will be affected by gravity and other types of fields created by these energies.

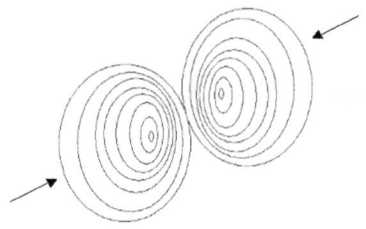

Attraction of Particles

Although all particles may be drawn together in this matter, as seen previously, particles with like spins repel and particles of opposite spins attract. Historically, three subatomic particles are explained in modern physics, electrons, protons and neutrons, in reality only two can exist on their own as stable particles, the proton and the electron since the neutron isolated alone will break down into a proton and electron within 15 minutes.

Studies have shown the masses of the particles to be:

Proton $= 1.6726*10^{-27}$

Neutron $= 1.6749*10^{-27}$

Electron $= 0.00091*10^{-27}$

What is interesting is that the mass of the electron and the proton added together is $= 1.6735*10^{-27}$, very close to the mass of the neutron $= 1.6749*10^{-27}$, a difference of $0.0014*10^{-27}$, which ends up to be an additional particle or particles with 1.53 times the size of an electron.

Under extreme conditions during the creation of our universe, high energy levels and atomic pressures turned excess protons in a nucleus into neutrons through the capture of electrons. These combinations of opposite spin particles, not unlike the hydrogen atom, created a neutral particle in the nucleus of the atom helping balance the strong repelling energies of proton, which have same spins and charges.

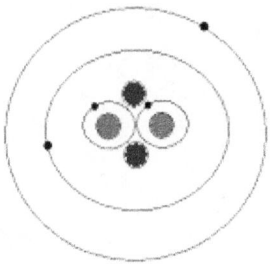

Neutrons in the Nucleus

In summary, a particle can also be viewed as a shockwave that is manifested as a larger single entity created by energy moving through a medium made of fundamental particles. The density and the energy (temperature) of these particles determine the velocity of this shockwave. This wave, similar to sound waves, can only exist at one velocity and cannot go any faster or slower in the material it is passing through. As the density or the temperature of this material changes, so will the shockwaves velocity – the speed of light.

The Photon

Light and other forms of electromagnetic radiation are believed by the modern scientific community to be the bombardment of particles called "photons" or what Einstein called "light quanta". These are particles that are believed to be ejected from the orbits of electrons, moving in a straight line and traveling at the speed of light until being absorbed by another object. Photons, having almost no mass, still have a noticeable displacement of material upon impact raising some questions. Einstein states that the faster a particle moves that it's mass increase until at the speed of light it has infinite mass. How then can anything travel at the speed of light and have mass?

In the "Double Slit" experiment performed by Italian mathematician and physicist Francesco Maria Grimaldi (1816-1863), a phenomenon called optical interference was created which can only occur if light is moving in waves (Ditchburn, R.W., pg. 119). In this experiment (also performed by Thomas Young), light from a single source is shone through a single slit in a barrier then continues through a second barrier with Double Slits. Light waves bend or diffract (breaking into pieces in Latin) as they pass through the slits causing the light waves to curve into what is called spherical waves. As light from both slits hits a viewing screen, the wave peaks and valleys converge at the same points on the screen creating dark areas and light areas of contrast called an interference pattern. Waves that are in phase create constructive interference and brighten the area in which they strike. Waves out of phase cause destructive interference and cancel each other out causing the darker areas on the screen. This optical interference can only occur with light that is correlated or coherent, coming from the same light source (Ditchburn, R.W., pg. 116). Monochromatic light, meaning "one color," is more suitable for this type of experiment as only one wavelength is produced creating a more vivid interference pattern; white light is a mixture of all wavelengths and produces a less visible pattern. Similar interference patterns can be also seen with the interferometer that has been adjusted so that the two light paths are of different lengths causing them to arrive at the target out of phase.

Double Slit Experiment

What causes these waves to **diffract** or bend as they pass through the slits? This bending of light waves evidently is due to diffraction from an interaction with the barrier as the light passes through the slits. The closer the light waves come to the barrier material, the sharper the bend.

The bending of light waves normally occurs when light moves from one medium into another and changes velocity, for example, light transitioning from air into glass. Light travels through air at approximately 185,226 miles per second, or 99% the speed of light in a vacuum. Light passing through glass however, only travels about 114, 989 miles per second, or 62% the speed of light. As the light transitions from air into glass, the light waves slow down and the wave front changes direction causing the light waves to bend or **refract** as it passes through boundary layers of different medium.

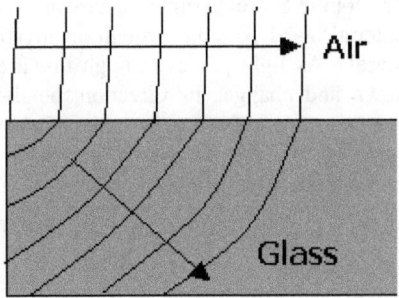

Refraction in Glass

A similar process called refraction can be seen with a spoon in a glass of water. Light waves transitioning from water into air after reflecting off the spoon speeds up and changes direction creating the illusion of a bent spoon. The refraction index varies from material to material based on the Optical Density of the material. The denser the optical density, the more the light bends as it passes from one material into another. Optical density is not the same as mass density, however there is a correlation but not linear.

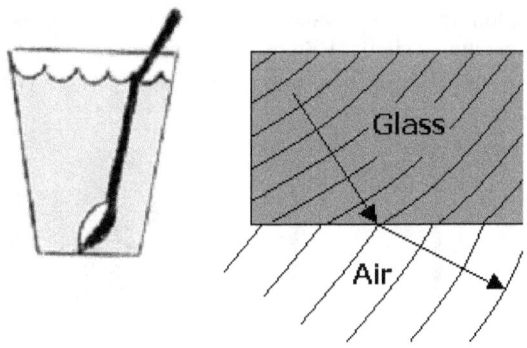

Refraction in Water

The light in the Double Slit experiment bends more the closer it is to the barrier material, as if passing through an invisible field of higher density material. As light passes through this area of higher density it slows down and changes its direction, bending outward. The combined wave becomes curved as a spherical wave; the two spherical waves created by the two slits are what cause the interference patterns on the target surface.

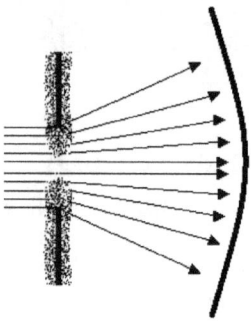

Slit Experiment Diffraction

In 1909 physicist Geoffrey Taylor performed a similar test using dark filters to block most of the photons, allowing only one photon at a time to pass through the slits. Over time the patterns of interference reappeared, showing that single photons will align themselves in a pattern of crests and troughs indicating that the single photon travels as a wave. Electrons also have been fired one at a time creating a similar pattern, refracting as it passes through the slit and falling on the target in crest and trough patterns (Ditchburn, pg. 643).

One at a Time

It is believed that the only way this could happen with single particles is if each photon or electron is divided and passes through both slits at the same time (Hawking, pg. 61). How else can crests and troughs be created? These patterns rely on interference, the interaction between two waves simultaneously, that create the constructive or destructive patterns on the target, light and dark spots. Otherwise an individual electron or photon would have to exist and then not exist; an on then off switching as it moves though space in order to produce this pattern by its self. For example, when the photon strikes the target when "turned off", it would create a light spot and when striking the target when "turned on", it would create a dark spot. Over time these will form interference patterns on the target. These light and dark spots stay in the same areas, and although created by moving particles, create a **standing wave** where the visible wave seems to be stationary.

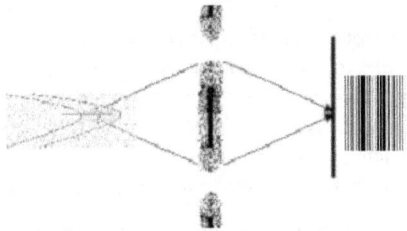

One Photon Two Slits

Let's assume again that a medium sea exists as if a field around all particles and that this medium moves into these particles through a cooling or condensation process creating varying densities in the medium. Matter attracts matter; therefore, the medium closest to a large mass such as the sun will be denser than the medium out in empty space. As light energy is passed from one particle of the medium to another in waves of various wavelengths, it refracts slightly as it transitions through denser medium, changing the direction it is traveling. Further away in the less dense medium, light waves travel in a relatively straight line. In the Double Slit experiment, light waves passing closest to the walls of the slit curved as they passed through slightly denser medium creating a spherical wave moving towards the experiments target. These spherical waves from both slits created the interference on the target.

Einstein saw a similar event as light waves bend as they pass through the suns strong gravitational field changing the apparent position of a star. This proved to him that light was composed of matter – a particle that is now called the photon. Was this gravity that bent the light waves or the change in the aether's density close to the sun? Possibly they are the same thing.

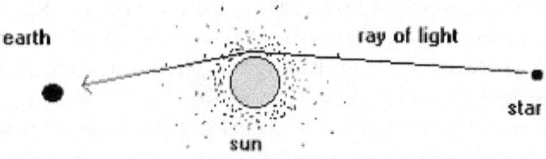

Light Rays

The flow of electrons also affects this medium that surrounds all things, generating electromagnetic waves in various wavelengths. An electron transfers electrical energy at the speed of light but does not travel at the speed of light itself; much like air or water molecules that oscillate moving particles at the speed of sound in the form of waves do not themselves move at the speed of sound.

Energy transferred from one electron to another in a wire carrying alternating current moves back and forth at the speed of light at a specific rate (the electrons themselves remain localized in the wire). The field surrounding the moving electrons generates a disturbance in the form of a wave that oscillates back and forth outside of the wire and is dispersed out into space in all directions as electromagnetic energy known as radio waves. This wave must have something waving much like air or water particles do.

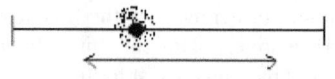

Electron in a Wire

The electrons themselves can stay relatively confined and only pass the energy from one electron to the other at the speed of light. Electrons generate electromagnetic radiation by jumping "back and forth" from one shell of an atom to another while orbiting around atoms or when jumping from one atom to another. Notice the "back and forth" motion? Remember, the "five ball pendulum", where there were five or so balls hanging on strings lined up in a row next to each other? Energy is transferred from one ball to another with very little movement of each ball; only the first and last balls really move at all. The same is true with electrons and the transfer of their energy and may also be true for energy traveling as light waves through this medium that envelopes all things.

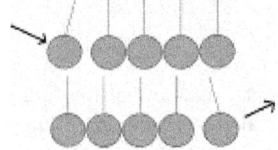

Five Ball Pendulum

Wave patterns not only are evident in the fore and aft movement in the direction of propagation, but in all directions, which bring us to the phenomenon of polarization. All particles find an equilibrium or neutral state when they expel all energy and come to rest. As energy passes through these particles however, they oscillate back and forth like pendulums, up and down, left and right or fore and aft. This back and forth waving of energy is another indicator that light moves in waves.

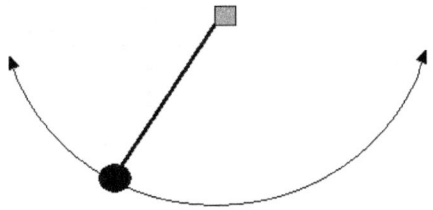

Pendulum

In the electromagnetic realm, this direction of wave motion is called polarity and is evident in all wavelengths of electromagnetic energy. The use of polarized sunglasses and the vertical or horizontal positioning of radio antennas are all examples of common utilization of the polarity phenomenon.

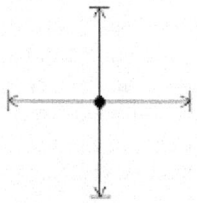

Wave Dimension

For example as light waves travel through certain crystals such as tourmaline, the crystal formations form tunnels that only certain wavelengths of light with a specific polarity will pass (Polkinghorne, pg. 18).

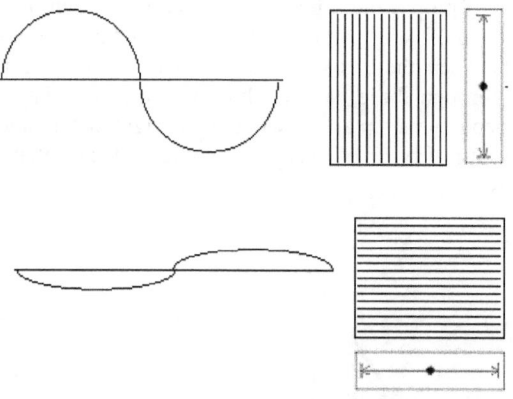

Vertical and Horizontal Polarity

Waves with opposite polarities become blocked such as vertically polarized waves trying to pass through a horizontally polarized pane of glass.

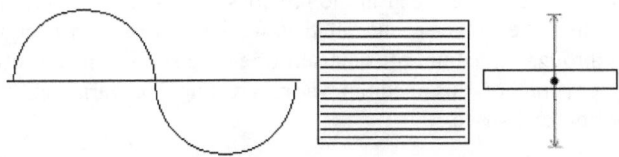

Opposite Polarity

All materials molecularly are constructed differently and have various responses to the electromagnetic energy. Some reflect, some absorb and some pass the energy straight through them. When absorbed, most of the radiation is then re-radiated back out at other wavelengths such as infrared radiation.

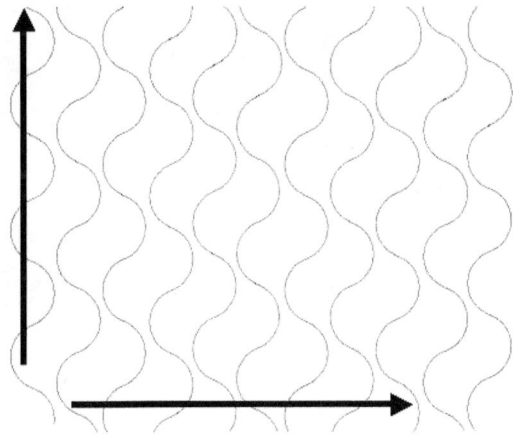

Electromagnetic Tranverse Waves

Particles move back to their natural state passing energy to the next particle in waves called plane waves or linear waves. Some waves not only move energy in the direction of the wave but also perpendicular to the wave in what is known as a transverse wave – these types of waves are only possible in **solids**. Energy in the electromagnetic realm moves as transverse waves at the speed of light in all directions as seen in the previous diagram. This in turn suggests that the universe and all that we know exists as energy passing through a medium of some sort of elastic solid. The density and temperature of this medium varies creating the variations in gravity, time and space.

Waves or Particles

In the year 1900, as previously discussed, German physicist Max Planck hypothesized that energy emitted or absorbed by an atom can only be transferred in distinct quantities, which he called "quanta", Latin for "how much". This "quanta" soon played a critical role in the new study of quantum mechanics as it required that oscillating atoms only possess the energy values of E or a whole number multiples of that value. The formula for E = h X f, where Planck's constant "h" is multiplied by the frequency "f", of the oscillating body (Bord, pg. 374).

The rationale behind this formula is that one full oscillation of a particle; let's say an electron, during a specific time frame, would use a specific amount of energy. Doubling the oscillations during the same time period would take twice the amount of energy and tripling the oscillations would therefore take three times the energy and so on. This oscillation is the frequency at which the electron vibrates and as it turns out is directly proportional to wavelength of the electromagnetic energy waves such as light waves emitted from an oscillating electron, whether it is from a star or a light bulb.

Planck's constant, "h" is measured in Joules (symbol J), the International System of Units (SI) energy measurement for heat, electricity and work. Planck's constant, h = 6.63 X10^{-34} Joule seconds (Bord, pg. 375).

There are two oscillating bodies or particles that are commonly referred to as energy carriers in the quantum realm, the electron and the photon. Each of these particles can be scientifically considered as either particle or wave in nature as mathematical formulas have been developed that seem to indicate them as both.

The neutron, outside of the nucleus has a very short life span of only a few minutes, breaking down, believe it or not into a proton and an electron in a radioactive decay process known as beta decay. Even within the nucleus, this radioactive decay occurs naturally in some elements, as in carbon 14, where an electron is ejected from the nucleus at a high rate of speed and a neutron spontaneously converts into a proton; a smaller particle known as a neutrino is also emitted as a release of energy (Bord, 2000, p. 417). Carbon 14's nucleus contains 6 protons and 8 neutrons and converts into Nitrogen 14 with 7 protons and 7 neutrons when the neutron goes through beta decay and becomes a proton.

Many other short-lived elementary particles have been discovered in recent years, bringing to light the complicated constructs of the subatomic world. These include leptons and quarks, the building blocks of electrons and protons, as well as antiparticles, which mimic other atomic particles inversely as if viewed in a mirror having opposite charges and spins. The contact of a particle and anti particle results in the annihilation of both particles in a large burst of energy (Bord, 2000, p. 457).

Elementary particles are the building blocks of the universe and are the fundamental constitutes of matter, either bosons or fermions depending on the type of spin (up or down). Particles normally linked with the construction of matter are called fermions and have a half-integer spin (odd integer (number) spin $+1/2$). Fermions are divided up into twelve categories called "flavors" and three other groups called "colors" (Bord, 2000, p. 479). Particles associated with fundamental forces - the strong, weak and electromagnetic and possibly gravitational forces are called bosons, named for Satyendra Nath Bose, having an integer or whole number spin (Bord, 2000, p. 458).

Fermions:

1) **Quarks** - up, down, charm, strange, top, bottom

2) **Leptons** - electron neutrino, electron, muon, muon-neutrino, tau and tau-neutrino

Bosons:

1) **Gauge bosons** - gluon, W and Z bosons and photon

2) **Higgs boson** and the theoretical **graviton**

Only four charged subatomic particles exist that are stable particles, the electron, proton, positron and antiproton (Asimov, Isaac, pg. 246). About 500 other particles have been identified in experiments using particle accelerators in recent years, all of which are unstable and exist only for short periods of time. Subatomic particles are composite particles created out of fundamental particles such as quarks and gluons that have no known substructure. These particles can only exist under certain conditions, known as "confinement" such as in the neutron, which decays into a proton and an electron in a matter of minutes outside of an atoms nucleus (Bord, pg. 417). Fundamental particles such as leptons and quarks are highly unstable and can only exist in confinement, attached in pairs or triplets by the "strong force"; these elementary particles in turn are the main components of the electron (a lepton), proton (two up quarks and one down quark) and of course the neutron.

Although these particles may seem to exist as individual masses, they actually may simply exist as waves in the field of fundamental particles that fill the vastness of empty space. As mentioned earlier, subatomic particles possibly only exist as a quantity of these smaller fundamental particles under certain conditions. As energy moves through these particles they temporally coalesce in a form of phase transition, passing energy from one particle to another in much the same way as do fans executing the "wave" in the stands of a sports event. Since energy cannot be created or destroyed (Law of the Conservation of Energy), this energy continues on in waves from particle to particle. This energy at the quantum level in the form of Plank's Constant also cannot be divided, transferring all of its energy to the next particles.

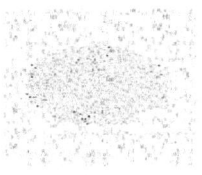

Electron as Particles Temporally Coalesced

Under specific conditions, this pressure wave possibly becomes trapped in a continuous circular motion, confined to a specific pressure gradient around an atom where it alone can exist, manifesting itself as an electron. As energy is absorbed into this wave, the wave changes momentum and instantaneously repositions itself in a pressure gradient where different conditions allow the "particle" to reappear as an electron in another shell or pressure gradient; such as transitioning from orbit 5 to 8 at a new velocity. Think of this as fans performing the wave around the stadium only in row 55, then halfway around stopping and shifting instantly to only row 75 for the rest of the way around. Although there are people in every seat, the wave manifests itself as people stand and raise their hands one after another and then sit down in a continuous fashion. The distance the wave has to travel as it circles the stadium varies from row to row as well time it takes to complete one circle.

Transition from Orbit 5 to 8

If we look at the electron as a single particle in orbit, as if a satellite around the atomic nuclei, the electrons wavelength of the smallest orbit of the hydrogen atom with a velocity of the $2.19*10^6$ m/s (meters per second) is: (λ is wavelength)

λ = h/mv (Planck's Constant and Momentum), then:

λ = 6.63 X 10^{-34} J-s/$1.995*10^{-24}$ kilogram-meters per sec.

or:

λ = 3.32 X 10^{-10} meters or .332 nm (nanometers)

Nanometer = one billionth of a meter (0.000000001)

If r is the radius of an orbit then the circumference of that orbit is = $2\pi r$, which is also its wavelength.

Since $2\pi r = \lambda$ = .332 nm then the radius of the smallest orbit or orbit **1** is:

1 radius = $\dfrac{.332 \text{ nm}}{2\pi}$ = $\dfrac{.332 \text{ nm}}{2 \text{ X } 3.14}$ = $\dfrac{.332 \text{ nm}}{6.28}$ or .0529 nm

Orbit **2** would **theoretically** then be **2λ** or

= 2 X .332 nm = .664 nm.

The circumference of the next orbit or orbit **2** is:

2 radius = $\dfrac{.664 \text{ nm}}{2\pi}$ = $\dfrac{.664 \text{ nm}}{2 \text{ X } 3.14}$ = $\dfrac{.664 \text{ nm}}{6.28}$ or .106 nm

Orbit **3** would then be **3λ** or = 3 X .332 nm = .996 nm.

The circumference of the next orbit or orbit **2** is:

3 radius = $\dfrac{.996 \text{ nm}}{2\pi}$ = $\dfrac{.664 \text{ nm}}{2 \text{ X } 3.14}$ = $\dfrac{.996 \text{ nm}}{6.28}$ or .159 nm

Orbit 1 radius = 1λ = .0529 nm

Orbit 2 radius = 2λ = .1057 nm which is 0.0529 nm further

Orbit 3 radius = 3λ = .1586 nm which is 0.0529 nm further

Orbit 4 radius = 4λ = .2115 nm which is 0.0529 nm further

Orbit 5 radius = 5λ = .2643 nm which is 0.0529 nm further

Energy = h (Planck's Constant) X Frequency

Frequency = Speed of light

Wavelength

Wavelength = Speed of light

Frequency

Speed of light = c   Frequency = f   Wavelength = λ

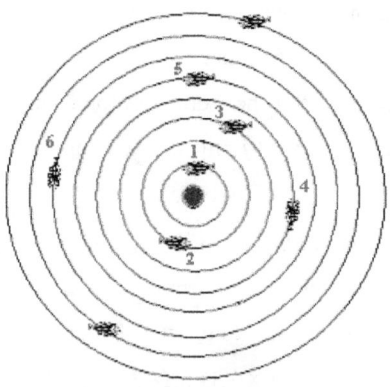

Waves of Particles

We see that if we increase each orbit by multiples of the wavelengths, each orbital radius is increased proportionately as in the previous illustration, 0.0529 nm further out for each new orbit. This is not the case however; each orbit changes according to variations in energy in fixed quantities of multiples of h, Planck's constant, not fixed distances from the nucleus. These orbits are not measured in radius or circumference, but in Electron Volts. Each orbits radius decreases slightly the further out in orbit the electrons are. As an electron transitions from one orbital or shell to another, the distance between orbits also changes.

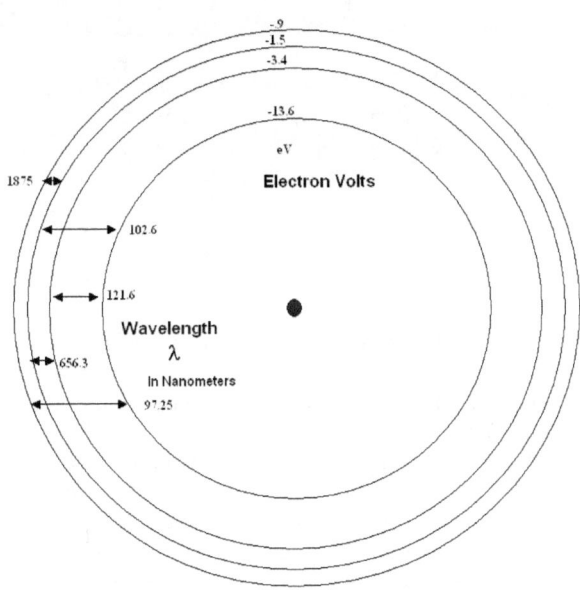

Radius Change in Orbit

The difference between orbit 1 and orbit 3 is greater than the difference between orbit 4 and orbit 6. The velocity of the electron also varies from orbit to orbit. As the distance of the electron from the nucleus increases, the force that pulls the electron towards the nucleus decreases by the square of the distance. This phenomenon is common to many processes such as the change in gravity, magnetism and the change in intensity of electromagnetic energy or light as the distance from the source increases. This in turn has a direct affect on the electrons velocity.

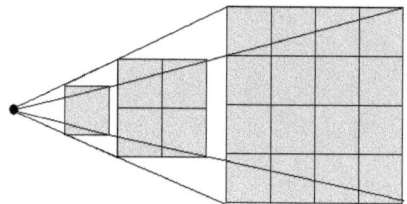

Attraction Decreases by the Square of the Distance

If an electron has the mass m, and the nucleus has mass M, and the electrons distance from the center of the nucleus is r, then g is the force that the nucleus exerts on the electron or $g = Gm\,M/r^2\,/m$ (where G is a universal constant for the attraction of mass). If the electron would fall towards the nucleus, the nucleus's attraction will cause it to accelerate toward the center of the nucleus. According to Newton's second law (F = ma), this acceleration g must equal:

$$g = (Gm\,M/r^2)/m, \text{ or } g = GM/r^2$$

For the electron as a single particle, the proportionate change in energy input does not produce a proportional change in orbit. The velocity of the electron varies according to the change in the nucleus's attraction due to the change of distance in radius from the nucleus.

The **angular momentum** (p) of each orbit or mrv is:

$$F = \frac{mv^2}{r} \qquad \frac{v^2}{r} = \frac{GM}{r^2}, \quad \text{or} \quad v = \sqrt{\frac{GM}{r}}$$

(**M**ass x **R**adius x **V**elocity)

$$mrv = \frac{h}{2\pi}$$

Therefore the angular momentum for:

Orbit **1** = **mrv** = $\underline{6.63 \times 10^{-34}\text{ J-s}}$

$$6.28$$

Planck's Constant can be looked at in two ways:

In Joule/ seconds = 6.63×10^{-34} J-s or

Electron Volts (eV) = 40136×10^{-15} eV/ Hz

Angular Momentum (mrv) therefore = 1.0552×10^{-34}

Velocity in the lowest orbit is 2,189,638 m/s

Mass of an electron of $9.11 * 10^{-31}$ Kg

Velocity in the lowest orbit is $2.19 * 10^{6}$ m/s.

Radius at lowest orbit = 5.29×10^{-11} meters

Wavelength = λ = .332 nanometers

Momentum (p) would be: 1.995×10^{-24}

The mrv = 1.0552×10^{-34}

If we go up to $2h$ or the next energy level up we get:

$$mrv = \frac{2h}{2\pi}$$

If the radius were **twice the ground state:**

Radius = r = 2 X 5.29 X 10^{-11} or **1.06 X 10^{-10}**

Velocity would be = **2,189,638 m/s**

Mass of an electron = **9.11*10^{-31} Kg**

Momentum would be = **1.995 X 10^{-24}**

Angular Momentum = **2.1104 X 10^{-34}**

If we stay at 2h:

$$mrv = \frac{2h}{2\pi}$$

And the radius was **three times the ground state:**

Radius = r = 3 X 5.29 X 10^{-11} or **1.59 X 10^{-10} (increased)**

Velocity would be = **1,459,758 m/s (decreased)**

Mass of an electron = **9. 11*10^{-31} Kg**

Momentum would be = **1. 330 X 10^{-24} (decreased)**

Angular Momentum = **2. 1104 X 10^{-34} (same)**

Looking at the electron as a particle, we see that at the same level of energy, 2h, we can have different velocities and different linear momentums. As far as the electron goes, the higher its orbit is the lower its velocity and the lower the linear momentum. Likewise, the lower the orbit of the electron, the higher its velocity is and the higher its linear momentum is. Angular Momentum (mrv) must therefore be calculated in order to show the total energy of the electron.

We also see that attraction forces change according to variations in orbits; the higher the orbit, the lower the attraction forces are. This change in distance has a direct effect on the velocity and linear momentum of an electron. Also, the higher the energy of the electron, the further out in orbit an electron will go; but the further out the electron is in orbit, the slower it travels and the less linear momentum it has. What then creates the fixed wavelengths of energy emitted from these systems? How do these slower electrons create the higher energy photons of shorter wavelengths? It seems the longer wavelength of the electron, the shorter the wavelength of the photon it creates – a paradox of sorts.

When sound is created from a guitar string, the string oscillates back and forth at a specific rate every second. Its wavelength is equal to the length the string physically moves, first one direction and then the other. The longer the wavelength of the strings oscillations, the lower the frequency is and the less energy is being used; the shorter the wavelength, the higher the frequency and the more energy used. To double the frequency of the strings oscillation is to double the energy radiated by the string at the same amplitude. This strings motion disturbs the air around it and the energy it radiates is dispersed into the air as compression waves or plane waves. Each of these plane waves travels away from the string at the speed of sound in that *medium* (air) at fixed wavelengths.

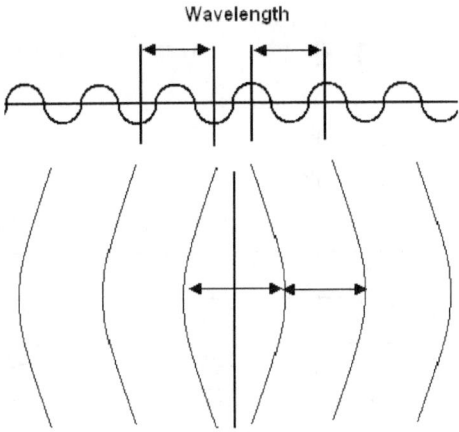

Guitar String Oscillations

An electron likewise oscillates back and forth as a guitar string. It does this in two ways, during its normal orbit around the nucleus of an atom in a circular motion at multiple levels and as it momentarily jumps from one orbit to another. The circular motion of the electron in orbit around the nucleus traveling a full 360 degrees is one full wavelength with each change in orbital radius a different wavelength. The velocity of the electron in a specific orbit produces a repetitive motion that repeats thousands of times per second, measured in Hertz (Hz). The faster the electron oscillates, and the smaller its orbit, the higher the frequency of the electron. The larger the orbit of the electron is the slower its velocity and the lower the frequency of the electron. The wavelength of the electron is dependent on its velocity and orbital circumference, not just the diameter of its orbit. One thing to keep in mind is that although an electron in a higher orbit is slower than lower orbiting electrons, it has **more energy** due to angular momentum. Faster electrons in lower orbit likewise have **less energy**.

Velocity in the lowest orbit is $2.19 * 10^6$ m/s

Radius at lowest orbit = 5.29×10^{-11} meters

Circumference = $2\pi r$ = or .332 nm

$$f = \frac{v}{\lambda} \qquad v = f\,\lambda$$

Orbit 1 = f = 6.59×10^{15}

Orbit 2 = f = 2.19×10^{15}

Orbit 3 = f = 1.098×10^{15}

Mass	9.11×10^{-31}	9.11×10^{-31}	9.11×10^{-31}
Radius	5.29×10^{-11}	5.29×10^{-11}	5.29×10^{-11}
Velocity	2,190,000	1,459,712	1,094,783
Wavelength	3.32212×10^{-10}	6.64424×10^{-10}	9.96636×10^{-10}
Hz	6.59218×10^{15}	2.19696×10^{15}	1.09848×10^{15}
mrv	1.0554×10^{-34}	1.40693×10^{-34}	1.58279×10^{-34}

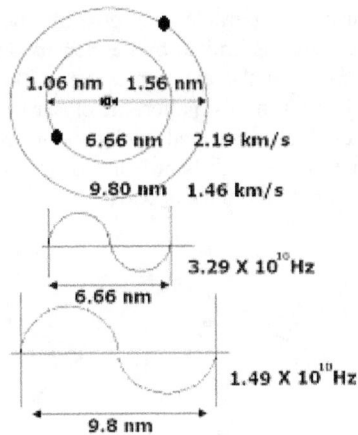

Wavelengths of Two Orbits

An electron also is constantly changing orbit as it absorbs energy bombarding it from all directions by photons. Each time the electron absorbs energy from a photon, it flies further out in orbit, and then releases the energy it absorbed and retreats back to a stable orbit. Each of these jumps or quantum leaps of the electron absorbs or emits a photon in fixed energy bundles previously discussed.

What is different here is that unlike the guitar strings wavelength, which is proportionate with the strings vibrations, the energy released from an electron works in an inverse way. The larger the movement of the electron, whether in orbit or during a quantum jump, the shorter the wavelength of energy is that is emitted. Likewise, the shorter the jump or smaller the orbit, the longer the wavelength of energy released. The answer to this dilemma may be in that the time for an electron to jump from here to there, no matter how many shells it may jump, is always the same; shell three to four or one to seven. If a constant say z for time is divided by the energy h that each shell releases (one packet for each shell jumped) then a three to four jump would be $z/1h$ the for the longest wave length of x and a one to seven jump would be $z/6h$ for a shorter wave length of y.

As covered earlier, a particle can possibly be viewed as a shockwave whose velocity is limited by the density and temperature of the material or *medium* the electron wave is flowing through. When viewing an electron in this matter, the closer to the nucleus an electron wave moves, the denser and cooler the *medium* is that the electron wave is moving through (since the particle itself may be a condensate of these cooled particles in wave form).

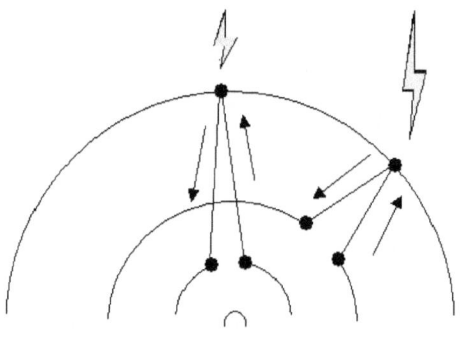

Quantum Leap

The further out that the electron wave moves from the nucleus, the hotter and less dense the *medium* theoretically becomes in which the wave is moving through. The electron as a fixed wave bundle of energy can only exist as a shock wave in one region at a time, at a specific velocity. If energy were added as a second fixed bundle of energy to the first wave (another photon wave is merged with the electron wave), the momentum would then be doubled ($2h$). With the momentum doubled but the mass of the combined particles in the wave still the same, the angular momentum and the orbit of the electron therefore must change. However, the present position of the wave in the *medium* it is presently moving through can no longer exist at this higher energy level, forcing the wave pattern to reform as a shockwave further away from the nucleus in a less dense region. This new wave then travels in a circular pattern around the atom further out staying within the new density bounds at a different velocity. This wave energy will be created at a higher angular velocity (mvr), which is actually a lower linear velocity but at a higher energy level. We must remember here that the electron itself

90

is not a particle moving the distance of low orbit to high orbit or vice versa, it is simply energy dissipating in one level and instantaneously reforming in another level.

We can again picture this as fans doing the wave around a stadium only in row 55. The more people in the row generating the wave, the longer it takes for the wave to move around the stadium. Inner rows have less distance to travel around the stadium but may have more people packed in the seats, slowing down the wave. Seats near the back of the stadium have many more seats for the wave to move through but possible has less people in the seats changing the velocity of the wave as well.

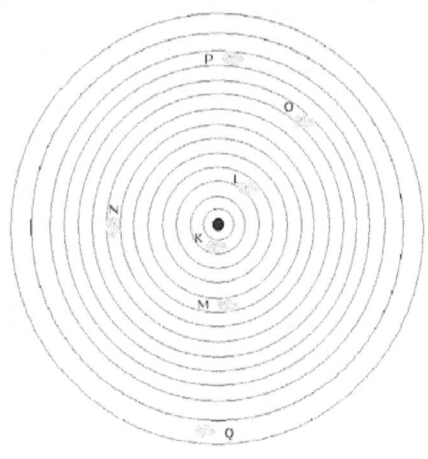

Temp in Kelvin	Velocity of Light
K 10.13	2,189,856
L 20.46	4,421,846
M 41.32	8,928,773
N 83.44	18,029,345
O 168.49	36,405,593
P 340.21	73,511,667
Q 686.97	148,437,772
∞ 1387.16	**299,731,634**

The previous illustration shows the seven shells that an electron wave can move through. The illustration shows how a possible temperature and density change of a *medium* from one shell to another affects the velocity of a shock wave through that *medium*. The velocity of an electron above Q, the highest orbit, breaks the bonds of the nucleus at the speed of light c, 300,000 km/sec, an escape velocity if you will, sending energy out through open space in the form of a photon in a straight line.

Sound

Generally speaking, sound travels slower in denser materials. The speed of sound within one material such as nitrogen or water however, is solely reliant on the temperature of that material, not the materials pressure or density. Sound travels faster however in solids and liquids than gases, because the molecular bonds of solid materials are much stronger than that of gases. Solids will have a higher sound speeds than liquids, and liquids will have a higher speed than gases (Bord, pg. 209). The speed of sound in air at standard temperature and pressure (25°C or 59 °F) is 343 meters per second (767 miles per hour). At 0° C or 32 °F the speed of sound in air is 330m/s (740 mph), 13 m/s slower. The speed of sound in water is about 1,500 meters per second (3,355 miles per hour) (Bord, pg. 224).

The wavelength of the sound wave at 240Hz is

λf, λ = v/f = (340m/s)/ (264 /s) = 1.29 meters

The speed of sound depends on the temperature, the density of the material (mass per volume) and the elasticity of the material or *medium* it is traveling through. As sound transitions from air to a solid, the speed of sound increases as well as its wavelength. As sound transitions from a solid to air, the speed of sound decreases as well as its wavelength.

Sound through <u>Air</u> - Wavelength 1.3792 Meters

Sound through <u>Cork</u> - Wavelength 2.0833 Meters

Sound through <u>Alcohol</u> - Wavelength 5.1667 Meters

(Sea Level)	M/Sec	Formula	MPH	Freq. Hz	W/L in Meters
Air (0° C. /32° F.)	331	0.44704	740.426	240	1.3792
Cork	500	0.44704	1,118.468	240	2.0833
Alcohol	1,240	0.44704	2,773.801	240	5.1667
Water	1,500	0.44704	3,355.404	240	6.2500
Brick	3,650	0.44704	8,164.817	240	15.2083
Wood (Oak)	3,850	0.44704	8,612.205	240	16.0417
Glass	4,540	0.44704	10,155.691	240	18.9167
Aluminum	5,000	0.44704	11,184.681	240	20.8333
Iron	5,103	0.44704	11,415.086	240	21.2625
Steel	5,200	0.44704	11,632.069	240	21.6667

0° C the speed of sound is 330m/s or about 745
Frequency - 240 Hz

Cork
1118.5 mph

Air
330 m/s or 740 mph

Air
330 m/s or 740 mph

Wavelength
1.3792 Meters

Wavelength
1.3792 Meters

Wavelength
2.0833 Meters

Alcohol
2773.8 mph

Air
330 m/s or 740 mph

Air
330 m/s or 740 mph

Wavelength
1.3792 Meters

Wavelength
1.3792 Meters

Wavelength
5.1667 Meters

Light

As light transitions from a vacuum to a new *medium*, or from less dense *medium* to a greater density *medium*, the speed of light decreases and its wavelength also decreases.

As light transitions from a *medium* to a vacuum or to a less dense *medium*, the speed of light increases as well as its wavelength.

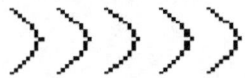

Light through <u>Air</u> - Speed 185,226.52 Miles/ Sec.

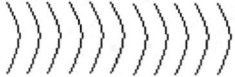

Light through <u>Water</u> - Speed 139,746.73 Miles/ Sec.

Light through <u>Crown Glass</u> - Speed 114,989.13 Miles/ Sec.

Medium	Speed of Light	Frequency in Hz	Wavelength in Nanometers	Percent of Light
Vacuum	186,282.40	4×10^{14}	680.00	100.00%
Air	185,226.52	4×10^{14}	676.15	99.43%
Water	139,746.73	4×10^{14}	510.13	75.45%
Glass	114,989.13	4×10^{14}	419.75	62.08%

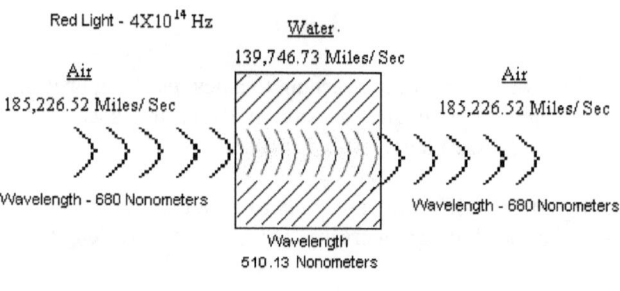

Red Light - 4×10^{14} Hz

Water·
139,746.73 Miles/Sec

Air
185,226.52 Miles/Sec

Air
185,226.52 Miles/Sec

Wavelength - 680 Nonometers

Wavelength - 680 Nonometers

Wavelength
510.13 Nonometers

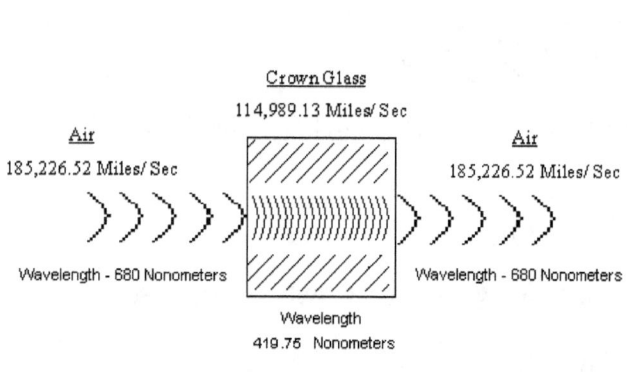

Crown Glass
114,989.13 Miles/Sec

Air
185,226.52 Miles/Sec

Air
185,226.52 Miles/Sec

Wavelength - 680 Nonometers

Wavelength - 680 Nonometers

Wavelength
419.75 Nonometers

Electromagnetic Waves

In the vacuum of space, absent of any interference of fields, the velocity of electromagnetic waves as energy is said to be a constant 300,000 km/sec or 186,000 miles/sec. In denser materials such as water and glass, the propagation is slower.

The simplest way to envision electromagnetic energy is to see how radio waves are propagated. If the length of an unshielded wire is proportionate to the wavelength of the cycle of electrons in an alternating current, electrons in a wire influenced by this electric current will generate an electromagnetic wave radiating out in all directions. Heinrich Hertz was the first discovered radio waves in 1888 (Asimov, Isaac, pg. 51)

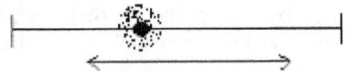

Alternating Current

An alternating current is a flow of electrons in a conductor whose current alternates back and forth, positive then negative, at a specific frequency or rate per second. The frequency of the electromagnetic energy or field generated by this current directly corresponds with the frequency of the alternating current. This shifting of the electromagnetic field can be sensed for miles depending on the strength of propagation and is dispersed as an oscillating electromagnetic wave in all directions. This energy possibly begins with the field of particles surrounding the electrons moving in the wire, which in turn disturb the particles in empty space around the wire as the electrons oscillates back and forth. Electromagnetic energy would then be transferred from one particle to another in this field of particles – which we will call the *medium*. This field of particles or *medium* is what will do the waving and possibly is the substance of which all things are made. The velocity of this transfer of energy depends on the density of the *medium*, 300,000 km/sec (186,000 m/sec) in a vacuum for example. All electromagnetic energy would then be the back and forth oscillation of this *medium* as waves at different frequencies.

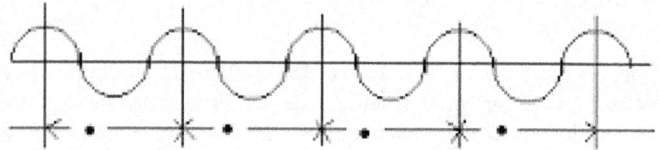

Wave Patterns

The transfer of this energy is at the speed of light. The wavelength is determined by the frequency of the oscillations, the higher the frequency the shorter the wavelength. The sine wave shown above depicts the electric current flow with the solid line representing no current flow, rising to a peak voltage or amplitude and decreasing back to zero volts. The sine wave then progresses below the line, representing the electrons in the wire flowing in the opposite direction to peak amplitude and then back to zero. These cycles continue thousands or even millions of times per second; electrons first flowing one way, decreasing in amplitude/ energy and then back in the other direction to full amplitude. The wavelength of electromagnetic radiation is measured as the distance between to peaks or troughs (valleys) in the cycle. These frequencies range from very low radio frequencies, to visible light, to x-rays and gamma radiation.

Some waves such as in radio waves are transmitted in half waves in one direction as in pulse and single sideband radio. This may be similar in all wavelengths of the electromagnetic spectrum in various transmission forms, both naturally and manmade.

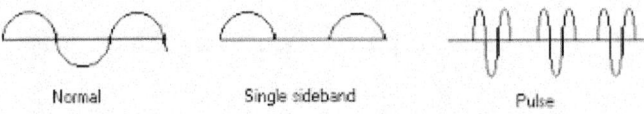

Normal   Single sideband   Pulse

Radio Transmission Patterns

With the fore and aft motion of the *medium*, we see that combining waves in phase, or combining waves out of phase, may interfere with their propagation.

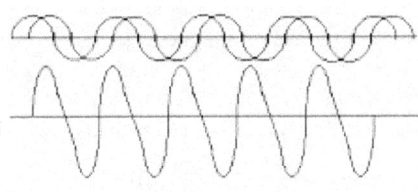

Combined Waves

As seen above, two peaks combined in phase amplify the peaks and the troughs.

Two peaks combined 180 degrees out of phase; cancel each other out leaving a flat line.

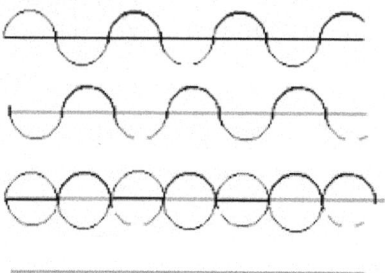

Cancelled Waves

This phenomenon exists in all ranges of the electromagnetic spectrum as well as sound waves in air and also with waves in water. Different materials react to different wavelengths of the spectrum, such as wave-guides (microwave antennas), radio antennas, photographic paper, x-ray sensitive film, and so on. From these materials we have created many different ways to detect and produce various wavelengths of electromagnetic energy for all sorts of things such as manufacturing, communication and recreation.

Doppler Effect in the electromagnetic spectrum is a shift in the frequency produced or reflected by moving objects, similar to the very noticeable change in sound waves as earlier noted with Einstein's train. With sound, as an object such as a train moves towards an observer, the sound the object produces seems to change in pitch as it passes. The frequency or pitch emitted from the object however is constant but is perceived to be higher as it approaches the observer. The movement of the object compresses the waves transmitted in front of it, causing a perceived change in pitch. Likewise, as the object passes the observer, the wave emitted from the object becomes stretched causing the observer to perceive the sound to be lower in pitch (Ditchburn, R.W., pg. 37). The same occurrence takes place with light waves and radio waves.

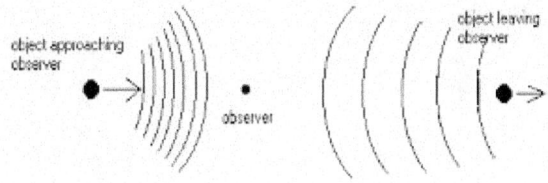

Doppler Effect

Technologies such as weather radar use this phenomenon to detect wind shear and possible tornado activity. Another use is the radar gun, utilized by police officers to detect speeders on the highway. Gravitational forces or movement of objects that emit or reflect light as covered earlier, produce red shift and blue shift in light waves.

The Electromagnetic Spectrum

Imagine the *medium* as a fine gas like particle that fills all things, surrounds all things and possibly is what all things are composed of. In the electromagnetic spectrum, an emitter, possibly a star, radio antenna or light bulb, stimulates or disturbs the *medium* and energy is passed from one particle of the *medium* to another. The velocity of the transfer of energy through the *medium* is relatively constant, and dependent only on the density and temperature of the *medium*. No matter what the velocity of the emitter toward the receiver or the velocity of the receiver towards the emitter, the velocity of the energy through the stationary space (*medium*) between the objects is always the same.

Even though the transfer of the energy through the *medium* or space is constant, the Doppler effect is still noticeable. If either the receiver or emitter or both move away from each other, the wavelength is stretched and perceived to be longer and a change in the light analyzed shifts towards red light, known as red shift.

Red Shift

If either the receiver or emitter or both move towards each other, the wavelength of the light will be compressed and perceived to be shorter. Within the light spectrum, a change in color is noticed, in what is known as blue shift. In any case, the wavelength emitted is actually the same. It is either that the wave as a whole is compressed or stretched by the movement of the emitter or receiver that causes the perception of waves to be compressed or stretched.

Blue Shift

As mentioned before, the velocity of the transfer of energy seems to be the same from one particle of the *medium* to another. The denser or closer together the particles of the *medium* are together, the longer it seems to take for energy to be transferred.

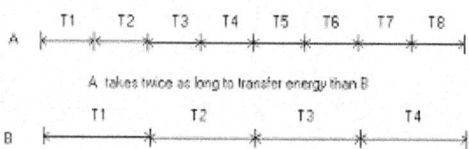

Velocity Through the *Medium*

As reviewed previously, energy is expressed in the quantum world in "packets" or "quanta" which means in Latin, "how much." This is also known as "Planck's Constant" with the formula $E = hf$ where E equals total energy, "h" is Planck's constant and "f" is the oscillation or frequency of the electromagnetic energy (Polkinghorne, p. 6). In the figure above, we show that it takes "T" amount of time to move one particle of the *medium* by one "packet" of energy into the field of the next particle, causing a chain reaction. *All* energy is passed from one particle to the next, developing a continuous chain reaction; this "packet" of energy cannot be split up. According to the "law of conservation of energy", no energy can be gained or lost. This is similar with electrons, as we will soon see.

Let's presume that each particle of the *medium* has a specific mass and inertia. It takes a specific amount of energy and time to move that particle from at rest to motion. Likewise, the *medium* particles take the same amount of energy and time to be returned back to rest. The higher the frequency, the more energy needed to start and stop each particle in a given period of time.

The question now arises as to how the velocity of light always remains the same to the observer. Light is always measured to be 300,000-km/ sec or 186,000 m/sec in a vacuum. How can this be? Einstein's whole basis for the theory of relativity is based on this discovery. He hypothesized that an increased velocity shifts space and time. As velocity increases, a clock starts to change time and mass begins to accumulate.

This hypothesis generates many scenarios, most of which possibly are only perceptions of the observer and not reality, similar to what we saw with the train and the embankment. For example a clock on a rocket launched from Earth may seem to be ticking slower to an observer. This may only be perceived because of the rate of speed, i.e. the Doppler Effect. Likewise, as the rocket returns, the clock may be perceived as ticking faster because of the Doppler Effect, but in reality, no time was lost or gained on the trip. As we will see, it may not be velocity at all that changes time but the acceleration and deceleration that affects an object as it changes velocity. A sphere traveling close to the speed of light going from left to right of a stationary observer may be perceived by the observer to be compressed (Born, Max, pg.319). Light would be absorbed and reflected off of the leading edge of the sphere. The sphere actually will run into the light reflected off the leading edge and make it appear to be compressed. Does this affect its reality? No, only in the way it is perceived, just as the sea bottom is distorted through the waves to the viewer on a boat.

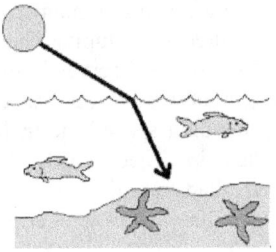

Distortion

Now let's look at the perceived speed of light to an observer. The *medium* in empty space is stationary and each particle is equidistant from each other, that is, in the natural "pressure" of the *medium*. Moving objects however and gravitational forces can compress or expand the *medium* causing higher and lower pressures or densities from its natural state. As an object moves toward a source of light at a constant velocity, the *medium* particles in line with both objects will compress. The number of particles will remain approximately the same between the emitter and the object, and only slowly change.

Imagine a tennis ball shot through a tube with approximately the same inside diameter as the diameter of the ball. Both ends of the tube are closed. The number of particles of air between the end of the tube B towards which the ball is moving, and the ball, remains the same except for a few that squeeze by the ball, with the air in the tube being compressed with increased pressure. Likewise between the aft end of the tube and the ball the number of particles remains the same even though it is being expanded and the pressure is decreasing.

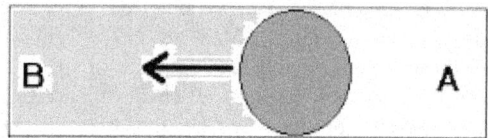

Compression of the Medium

Now imagine traveling towards Jupiter in a spaceship near the speed of light. Light emitted from Jupiter cannot travel faster than the speed of light relative to the observer in the spaceship. Just as with the ball in the tube, light waves would have of be compressed in front of the spaceship bending space-time in front of the ship and stretching out behind the ship. Theoretically, the spaceship could be much closer to Jupiter than would be perceived onboard. The question is what changes dimensions with speed? The spaceship or space itself?

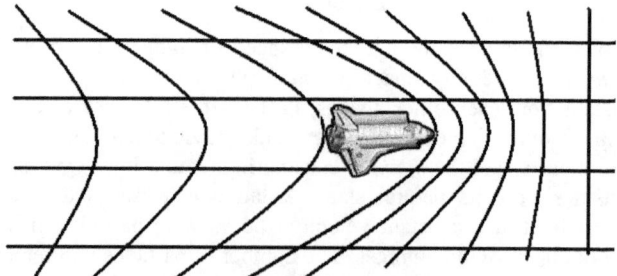

Compression of the Time and Space

Of course there is to be some difference between subatomic particles and sound through the air and water; this is because of the "law of conservation of energy". Energy transferred through air and water disperses energy into various factors, heat, motion, sound etc. In the subatomic electromagnetic spectrums, changes in energy only are seen in frequency changes of the emissions. The more energy released the higher the frequency of the radiating waves.

Now imagine all the particles of *medium* between the two moving objects. As the objects come closer, very few particles escape the pressure between the objects as it compresses. The number of particles increases in a given space as velocity increases. Therefore the speed of light changes proportionately to the change in velocity of the target to the emitter. This is relative to an outside observer. The outside observer cannot see this energy because he is outside the "compression zone". To the observer on the moving object, the speed of light has not changed no matter what his velocity. If the moving object was to stop, the pressure between the objects neutralizes and the speed of light still is measured at 300,000 km/sec. Now you say; if the speed of light is always the same to the moving observer, then there should be no apparent Doppler effect noticeable to the observer. Waves must then travel out in all directions from the source and not be affected by the compression. If this were true, the peaks of the electromagnetic waves moving towards an object would be perceived to be moving faster than light even though the light measured on the moving object is measured at the speed of light, 300,000 km/sec.

We see that the wavelength in miles is:

186,000 ...   or in ft ...   **983,559**

Frequency   **Frequency in kHz**

The wavelength is directly proportionate to the speed of emitter. How can you then have a Doppler effect with the perceived frequency of light shifting when the speed of light measured in the same wave doesn't change? This is a paradox.

One way of looking at this is to use a submarine traveling through water as an example. The sub sends a sonar beacon towards another sub in the water. The sub is traveling through the water at 30 knots. As soon as the sonar beacon exits the sub's transmitter, it moves in waves at a constant speed through the water **relative to the movement of the water particles** and **independent of the speed of the sub.** The speed of sound through water depends on the temperature of the water and is constant, in this case 3,355 miles per hour no matter what the speed of the sub. On the other hand, the faster the sub travels through the water, the shorter the wavelength and the higher the sonar frequency at the target due to the Doppler effect. The frequency of the sonar increases proportionately with the increase in velocity of the sub. The speed of sound through the water from the sub however **remains the same**. As you can see in this figure, the Doppler effect is easily noticeable.

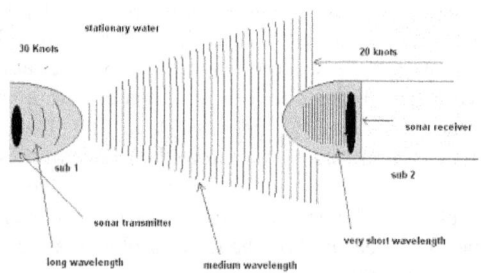

Sonar Waves

Fields and Forces

There are four fundamental interactive forces in nature. These four categories of forces studied in quantum mechanics are gravitational force, electromagnetic force, the weak force and the strong nuclear force; the weakest of these is gravity.

Gravitational force is the affect that every particle has on other particles and its force is directly proportional to the combined mass of objects within the field; this field's strength decreases by the square of the distance between the particles. For example, the sun and the earth: here the total gravitational attraction the sun has on the earth must be added to the gravitational pull the earth has on the sun. It is not just the powerful gravitational pull the sun alone that pulls on the earth but also the earths own pull on the sun. Likewise, a man standing on the earth not only is attracted by the earths pull on him but also his mass that is also attracting the earth. The greater the total mass (earth + sun), the greater the gravitational force that is pulling the objects together.

Long ago, Aristotle had reasoned, supposedly erroneously, that the velocity of falling objects were dependent on their weights. Galileo hundred of years later discover that the velocity of two objects of different weights fell at the same rate; this view is still held today. In fact, Aristotle was somewhat correct. A marbles rate of acceleration towards the earth is not only caused by the earth's attraction of the marble but also the marbles attraction of the earth. The acceleration rate towards the earth is due to the combined mass of both objects; therefore, a bowling ball will have an ever so slightly higher rate of acceleration. This is due to the fact that the combined mass of the earth and the mass of the bowling ball is slightly greater than that of the combined mass on the earth and the marble.

Gravity today is believed to be manifested by a mass-less particle called a graviton. This is done through an exchange of gravitons in gravitational waves between the particles of two bodies. This force, unlike the other three forces, only attracts and does not repel. Although it is the weakest of the four forces, its effects can be generated over a very large distance (Hawking, pg. 72).

The **electromagnetic force** is a force that is both attractive and repelling in nature and only acts between two or more electrically charged particles. Here, like charges repel and opposites attract, positive repels positive but attracted negative. Examples are where a proton attracts an electron but repel another proton; this is the attracting force that causes an electron to orbit the nucleus of an atom that contains the protons (Hawking, pg. 73).

As protons and electrons particles interact with each other, orbiting electrons will in certain situations be shared with the nucleus of other atoms; in a sense, tying the two atoms together in what is called a **covalent bond.**

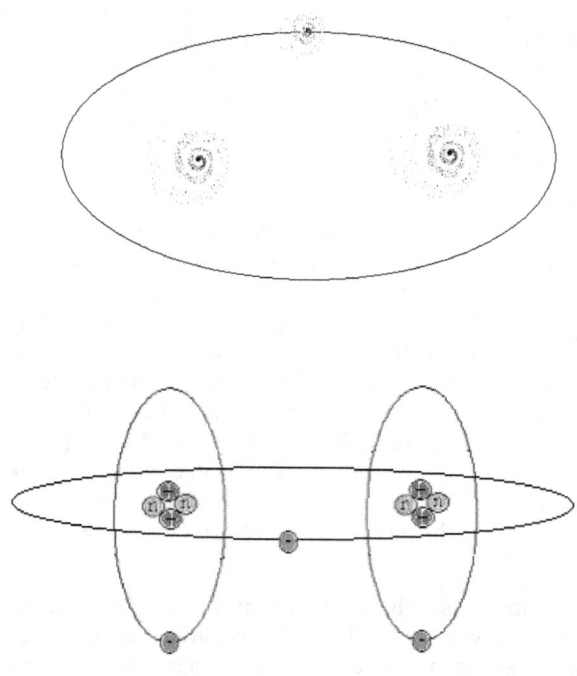

Bonding of Particles

In electrically balanced conditions, electrons and protons equal each other in numbers. In cases however where imbalances occur, electrons will try to passed from an area were excessive electrons are grouped to a region of electron deficiency. If a conductive path is available, a flow of electrons from one area to another in an electric current will occur– the principle of the battery. As these electrons move from one atom to another, a field is produced surrounding the electrons, moving with the electrical current disturbing space around them as they flow with what is commonly called an **electric field**.

Electron Flow

Magnetic fields are caused by the motion of electrons in a circuit as in the case of an electromagnet. Here the electrical field creates an attractive or repulsive force depending on the direction of the flow; its intensity is proportional to the amount of current in the circuit. A natural magnet such as magnetite also has a magnetic field but is caused by the alignment of atoms in the substance and possible flow of the electrons naturally within the substance. The synchronous movement of electrons moving in unison and perfect alignment of the electron orbits within the magnetic material produces these magnetic fields.

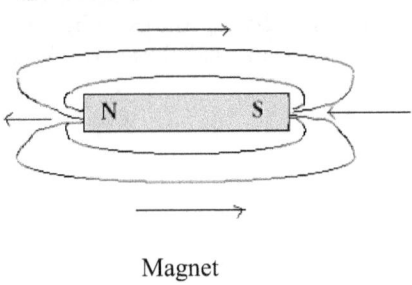

Magnet

The **electromagnetic field** in the form of a radio wave is also a variant of this electromagnetic force. Light waves are also electromagnetic waves that are at higher frequencies than radio waves and are manifested by the ejection of photons by the movements of electrons.

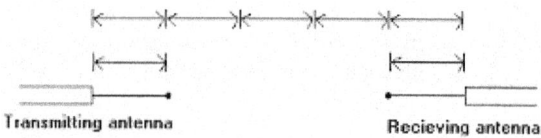

Transmitting antenna   Recieving antenna

Radio Waves

The **weak nuclear force** is found in nuclear radioactive decay such as beta decay, the break down of the neutron into an electron and a proton. Here, outside of the confines of the atom, a neutron no longer remains a neutron and breaks down due to the exchange of heavy W and Z Bosons. Particles react differently at different levels of energy, at specific energy levels symmetry can be broken causing particles to change, spontaneously breaking down (Hawking, pg. 74).

The **strong nuclear force** is the force that holds the building blocks of particles together. The quarks that make up protons and neutrons and the nucleus of the atom are possibly held together by the gluon. This gluon binds them together in a process called confinement, combining only specific combinations of different types of quarks (green, blue or red). Combining anti quarks and quarks may create the unstable meson of which then breaks down into an electron. The high energies of particle accelerators weaken the strong force, allowing these particles to be broken down into what appears to be "free" quarks and gluons (Hawking, pg. 76).

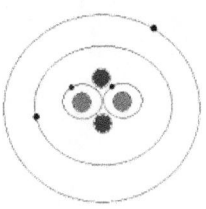

Atom

These forces are all forms of energy. Energy as we have learned is the movement or the potential movement of mass. Scientifically speaking, energy is a variable quantity that is a characteristic of an object. Energy is the ability to do work or that which is transferred when work is done (Bord, 2000, p. G-1). There are several ways to describe energy, kinetic, potential, gravitational, electromagnetic, chemical and nuclear. Energy can be transformed from one form of energy to another such as light into heat, but cannot be created nor destroyed. This is known as the **Law of the Conservation of Energy** (Bord, 2000, p. 98) written in **Sir Isaac Newton's** *"Philosophiae Naturalis Principia Mathematica,"* Latin for "Mathematical Principles of Natural Philosophy"- one of the most influential scientific works ever published.

As earlier discussed, there must be a *medium* or substance that all of these different forces travel through. It is the means of which all wavelengths of the electromagnetic spectrum are transferred. It is the guide wave for the particle world; photons, electrons, neutrons, and protons are all influenced by its existence or exist, as waves of it and it may also be the pressure or force of attraction called gravity. If energy exists as waves in empty space, what is it that is waving? In the sea, it is water molecules, in the air, air molecules. All particles may influence or even be composed of this *medium*. As we saw before with the radio antenna, the motion of the electrons moving through the wire transferred energy into the *medium* and back from the *medium* into the receiving antenna even through "empty" space.

Gravity

What is gravity? What is it that causes the acceleration of objects as they fall? Galileo determined that all objects, not accounting for air resistance, fall at the same rate at 9.8 meters per second per second no matter what size or weight on the earth.

Einstein noticed the similarity between gravity and acceleration and developed the "**Principle of Equivalence**", asserting that gravity and acceleration are the same. In a thought experiment, Einstein would imagine a man closed up in a in a "spacious chest" resembling a room. This room would be as if it were, in orbit with no gravity, when suddenly the room is accelerated by some force at a constant rate of 9.8 meters per second/ second, as if rocket engines were fitted on it and lit off. In this situation, not being able to see outside, the man perceives that this force is gravity, making gravity and acceleration indistinguishable (Einstein, 1952, pg. 67).

If gravity and acceleration are the same, how can we account for this "acceleration" in our universe? As a car takes off from a stand still and accelerates, the driver feels forced back into the seat. The higher rate of acceleration, the more the driver is pressed back. Now visualize the surface of the earth expanding outward at a rate of 9.8 meters per second, per second (per sec^2). All objects on the earth would maintain their weight similar to gravity. The problem with this however is that everything would also have to expand outward proportionally with the earth to stay in a balanced perspective and the rest of the universe likewise would have to expand at the same rate to stay in a proportional balance.

Now comes a dilemma; other objects with different gravitational forces must also be taken into consideration; for example, the moon, which has one-sixth, the gravitational force of the earth. With the earth's surface expanding outwardly at 9.8 meters per second per second, the moon's surface would be expanding only at 1.6 meters per second, per second (per sec^2). It wouldn't take very long at all to notice from an earth perspective that the moon would be shrinking very rapidly. In order to correct for this dilemma one would again have to change space and time to straighten this dimensional warping of space. The rate of lunar expansion therefore may be perceived to have sped up in order to keep a universe in balance or perceived distances would have to be greater to have this perception of no change.

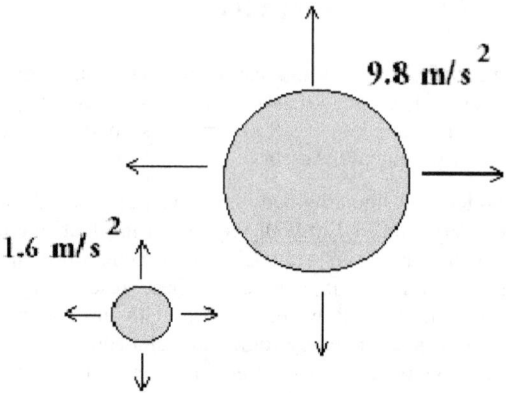

Different Rates of Expansion

This seems ridiculous. Again, we would have to resort back to the Lorentz Contraction Theory of dilated space and time in order to correct distorted space. However you look at it, all mass would eventually envelope all mass in this scenario, and everything would be annihilated; the earth would swallow up the moon and the sun, likewise the earth.

$$l'=l \sqrt{1-\frac{v^2}{c^2}}$$

Dilation of space

$$t'=\frac{t}{\sqrt{1-\frac{v^2}{c^2}}}$$

Dilation of time

Another way to look at this would be to look at the affects of deceleration on an object. As a car decelerates or slows down, instead of the driver being forced back in his seat as in acceleration, the driver is forced forward. If the man in the "spacious chest" were free falling, he would feel weightless as if in space. However, if the chest would start to decelerate due to some sort of resistance, such as entering into the earth's atmosphere, as more and more air resistance is generated on the chest, the chest starts to slow down and decelerate. The deceleration would be perceived as a slow increase in gravity within the chest; the greater the air resistance, the greater the deceleration and the greater perceived effect of gravity on the man in the chest.

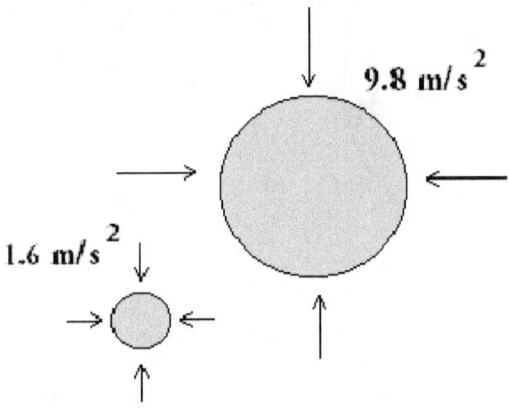

Different Rates of Contraction

If all objects in the universe were shrinking or slowly collapsing as the universe cools, all objects in space would then be slowly falling into themselves. As energy is released out into the universe from a particular cooling mass or object, the flow of energy could act as a wind, resisting the flowing particles of the cooling mass as they fall towards the center of the mass causing the particles to decelerate. The larger the object is or the closer to an objects core a particle comes, the more resistance to its movement. Likewise, the more resistance, the more deceleration and the more gravity there is that is perceived by the object.

Let's now look at another possible answer to this problem. If this gravitational force of acceleration is not caused by rapidly expanding mass, what else could it be?

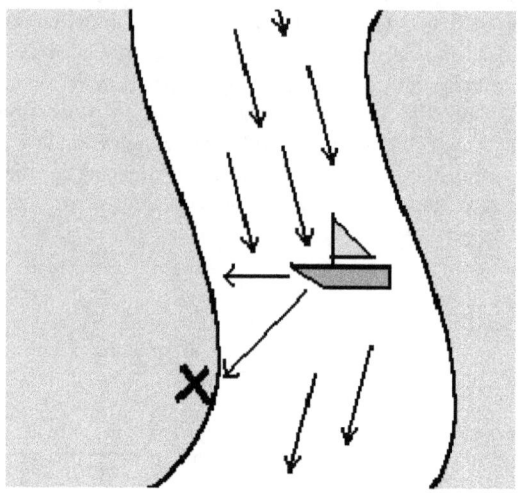

Flowing River

When a boat crosses a rapidly flowing river, the water current carries it along causing it to drift down stream. The same could be happening here with gravity. A rapid cooling of the universe after the big bang could be causing the condensation and rapid contraction of this medium, the aether cloud that surrounds all matter. As this material condenses, it forms in various levels of pressure, creating a "flow" of subatomic particles moving through and around all mass in its vicinity. This material would create regions of varying pressures; as the materials cool and condense the pressure drops causing higher pressures and energies to exist further away from the condensing mass. Higher pressures are outside of the condensing mass and lower pressures inside mass, this pulls in mass towards other mass, similar to wind on a sail.

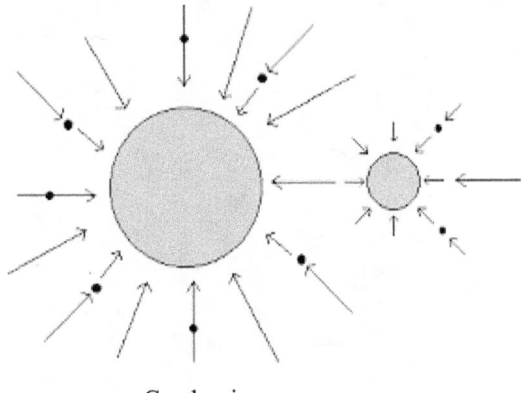

Condensing mass

The formula for the gravitational attraction of mass is:

$$F = \frac{M_1 M_2}{d^2}$$

Gravitational Attraction Formula

It may be that subatomic particles, electrons, protons and neutrons are simply clouds of this condensing spinning energy in the aether. The rate of spin would be proportional to its energy and the direction of spin would be its charge. Like charges would repel each other like spinning tops bouncing off each other. Opposite charges would attract each other spinning in opposite directions moving together like gears in a massive efficient machine.

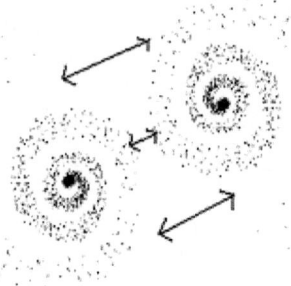

Like Spins Repel

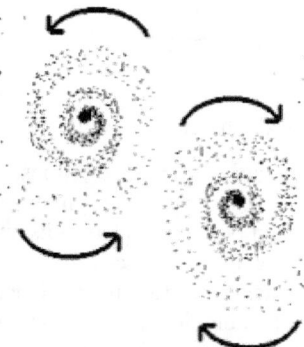

Opposite Spin Attract

Theoretically, it is these flowing and spinning clouds of aether energy, the same *medium* that transfers the waves of electromagnetic disturbances known as light, that cause gravity, electrostatic charge and magnetism. The particles of the aether do not necessarily have to flow; the "spin" may simply be energy waves trapped in orbit around these particles moving in different directions. These fields of "force at a distance" surround all mass and affect other mass in close proximity by its flow and spin without particles having to contact other mass. All of these forces decrease proportionally by the square of the distance from the mass-producing these fields. These fields easily radiate out into the open vacuum of space.

A boat crossing a river will move with the current, eventually moving at almost the same velocity of the water. The rate of acceleration depends on the boats mass and surface area to the force of the water. Unlike the force of the water on the side of the boat, the aether particles are so small that they exert a force on every particle throughout mass in general. Every atom, electron, proton, or neutron has a force on it from the pressures of the aether. Therefore since all mass is created of the same subatomic particles, all mass no matter what its density or size has the same gravitational force of 9.8 meters per second per second pressing on it here on earth.

A sailboat will accelerate as it is pushed by the wind. A sailboat however, can go as fast as three times the speed of the wind; how can this be? It is not the flow of the wind that moves the sailboat; it is the pressure differential on both sides of the sail that causes movement, in fact, a sailboat if sailed properly in a zigzag fashion, can sail into the wind. Likewise, it is not the flow of this aether that moves the object towards mass, it is the pressure differential between the higher pressures outside the central condensate particles and the lower pressures inside these central particles that cause mass to move together. The aether's differential pressure is proportional to the square of the distance to that mass, and the amount of mass. Differential pressure therefore increases proportionately closer to the attracting masses core an object comes. A large mass such as our sun will have a greater gravitational pull than our earth because the sun aether's has greater differential pressure.

Because of the earth's size, the majority of the aether flow (or pressure differential) moves through us and through all matter on the earth; every particle in every object is also absorbing this flow. Albert Michelson as we saw earlier designed a device he thought could measure the velocity of the earth as it moved through the aether – a device designed to detect aether flow, the Michelson Morley Interferometer.

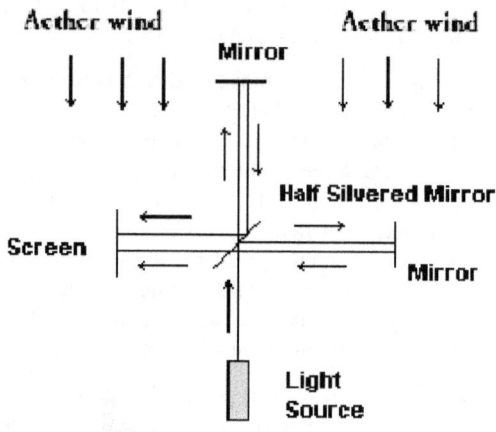

Michelson Morley Interferometer

As with the boat crossing the river, the current will drift the boat down stream, Michelson believed that an aether wind would drift light, causing his interferometer to display an interference pattern.

Interference Pattern

As the aether wind would flow across the light beam, it would slightly displace its reflection on the 45-degree half-silvered mirror. Half of the light would go through the mirror and reflect back again of another mirror. The aether wind would then slow down the light beamed into it. Both of these light beams would then be merged again at the silvered mirror toward an eyepiece. The slight change in distance that one light beam of light would travel (due to the aether wind) would cause an interference pattern in the eyepiece, as the light waves of both beams would now be out of phase. But this was not the case and the experiment failed. Why? Because they assumed that the aether flow was parallel to the earth's surface due to the earth's motion through space. In fact, this aether flow they were looking for may be perpendicular to the earth's surface flowing into it – known as gravity.

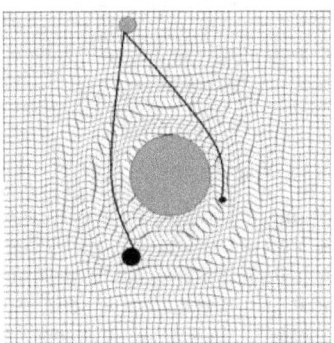

Radar Tracking of Mercury and Venus

Astronomer Irwin Shapiro's discovery in 1964 was in fact a modern day variation of the Michelson Morley Interferometer. Here, the beam was not light but a radar beam, and instead of reflecting off a mirror, the beam was reflected off the surface of the planets. Michelson and Morley were attempting to show that the light beam moving in the direction of the aether wind would be slowed down creating an interference pattern made by waves being out of phase. Shapiro showed that a shift in the radar frequency was created by radar energy being slowed down by the suns gravity. Here again, gravity is shown as the aether wind.

This is also evident during a solar eclipse when the corona of the sun can be seen around the shadow of the moon. This should not be the case. The size of the moon and its distance from the earth should block the suns corona totally. This was one of the first proofs that even the small gravitational field of the moon affects the movement of light around it. Starlight has also been known to bend or diffract by the suns gravitational pull, changing its perceived position in the sky. This was first seen in 1919 by astronomer Arthur Eddington while observing an eclipse off the coast of Africa (Isaacson, pg. 257).

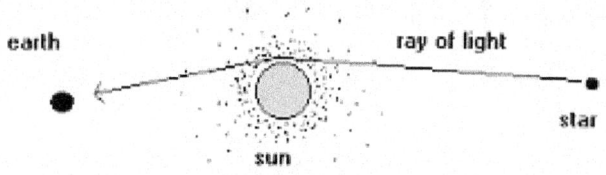

Diffraction of Light

At present, huge interferometers such as LIGO are being used to look for gravitational waves (Bartusiak, pg. 115). What are these gravitational waves? Could these waves flow through the aether, the substance that scientists are so reluctant to admit exists? Again, let's consider that all mass is slowly condensing material with the condensate material as the aether. As energy radiates out from the core, it cools and completes a phase change, turning it into an individual particle. As energy dissipates outward, the cooling material continues to condense, drawing more of this aether into it at a constant "flow". The closer to the core, the cooler this aether becomes, dropping in pressure similar to isobaric lines seen on weather maps depicting low pressure regions. High-pressure regions out side push material towards the low-pressure regions spiraling in a rotating motion.

This process over time has caused a cooling universe to condense or coalesce as mass attracts mass and starts to accumulate. If these different pressure regions did not generate during this process, the universe would stay a soup of thinly spaced particles, no light, just a dark cloud.

Over time, as mass attracts mass, large masses begin to accumulate more mass as their gravitational pull on smaller objects compounds dramatically. Now the large mass begins to heat up, squeezing out the energy it contains as if a sponge is being squeezed. Huge gravitational forces now cause the fusion of these fine particles into hydrogen and helium, the fuel sources for stars, and a star is born.

As the universe continues to cool, stars continue to grow until they run out of fuel. This fuel originally came from the dust and debris that the star pulled in through its strong gravitational field. Soon however, the space around the star cleared up, as if cleaned up by a gigantic vacuum cleaner and the star had to rely on the fuel it had stored within its self. Over billions and billions of years, this fuel slowly burned up, shedding its energy out into space, giving up much more energy than it absorbed. As its energy subsides, its gravitational force begins to overcome the energy supporting it from within. At first the star grows like a balloon in a desperate struggle to survive, but soon, with very little energy left to stay inflated can no longer support itself against its own gravitational field; it then contracts and collapses in on its self into a black hole. The gravitational force of its huge mass has overcome the nuclear fire which once raged within it, now nothing, not even light can escape this massive gravitational field. **John Michell** (1724-1793), English philosopher and author of *The Philosophical Transactions of the Royal Society of London* developed this theory back in 1783 (Hawking, pg.83).

Not all stars come to this so-called end. It all depends on the mass of the star, very large stars however may indeed become black holes, and others become white dwarfs, brown dwarfs, neutron stars or pulsars. When medium sized stars like our sun collapse, they explode in a supernova, blowing off its gaseous material out into space and become neutron stars. More massive stars over come by their own gravitational forces continue to collapse becoming black holes. Indian-born American astrophysicist and Nobel Prize winner **Subrahmanyan Chandrasekhar** (1910 - 1995) determined what is now known as the Chandrasekhar Limit, the threshold at where mass and energy levels allow for the development of a **Black Hole** (Hawking, pg. 85).

Black Hole's, once just a theory, are now popping up everywhere. It is believed now that every galaxy has one and may be what gave birth to the galaxies. As clouds of dust through gravitational forces pulled together, massive amounts of mass came together and developed mass beyond the Chandrasekhar Limit. Here mass levels were so high and gravitational forces so strong that energy levels within this congregating mass were too low to prevent it from collapsing and a black hole was born. Dust clouds in the close vicinity of the black hole started to spin at high speeds and the friction of the high velocity dust caused it to glow with enormous brightness, the creation of a quasar. The high energy and the enormous amount of mass being pulled towards the black hole generated the birth of stars, spinning at great velocities around the black hole. Dust and gasses near the center of this huge storm were slowly pulled into the black hole; the last light to escape glitters off the "event horizon", the last chance for energy to escape the massive gravitational pull of the black hole (Hawing, pg. 88).

Back to the scale of the stars, sun and planets this continues to be the case, causing mass to attract mass. The greater the mass is, the greater the attraction with a lower pressure; for example, between the sun and the earth; a higher pressure existing outside of both the earth and sun pushes them together. As light passes through this aether cloud surrounding a large mass, the density of the aether increases. The closer light comes to the large mass, the slower light travels, causing it to bend or refract, just as light bends while passing from air into water. The stronger this gravitational field, the denser the aether and the more light will bend.

Black holes may actually come in many sizes (Hawking, pg. 112). The Chandrasekhar Limit, although based on the amount of mass a non-rotating object can absorb before collapsing into a black hole, also must calculate the total energy of the object. For example, when a massive star runs out of fuel, at a specific energy level it can no longer support itself and will collapse. Mass is basically constant here and energy is the variable. Smaller stars may also become black holes in the depths of space where the stars mass can drop to temperatures close to zero Kelvin. Different amounts of mass therefore will become black holes at different temperatures.

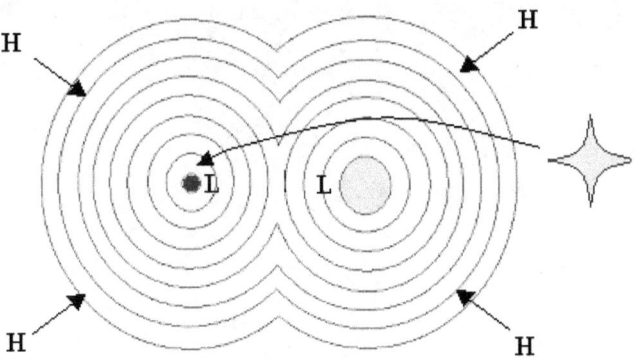

Isobaric Lines of Pressure

Although in today's scientific community, the concept of the aether is generally not accepted. Scientists however, do consider an invisible gravitational *field* to exist that surrounds all objects and increases in intensity with an increase in mass and decreases in intensity with distance way from the mass: the greater the gravitational intensity, the greater light is bent. What does this field of varying "pressures" consist of?

Magnetism

Magnetism is one manifestation of a field. An electromagnet is created with a coil of wire around a ferrous core material. When a battery is connected to the coil, an electron flow through the coil creates a field around the coils core called a magnetic field. The magnetic field contains lines of flux which seem to emanate from one end of the core, moving around to the other end, and can be seen by sprinkling iron filings on a piece of paper and putting a magnet under the paper.

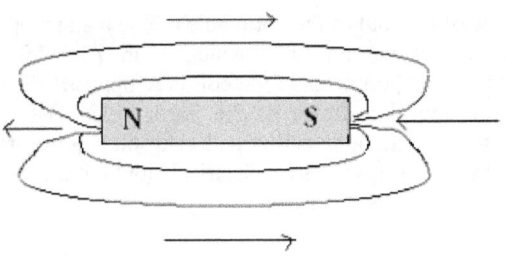

Magnet

What are these lines of force? In space or in a vacuum they exist but there supposedly is no mass associated with this flux. Like charges repel and opposites attract; when a north pole of a magnet is pushed toward another magnets north pole, the second magnet will move away. What *medium* transfers this energy?

As seen with the antenna, an alternating current induces a field, which flows with the current emanating out into space. With DC current in a wire however, current flows continuously from the negatively charged source of a battery towards the positive; with an electromagnet, the current flows from one end of a coil to the other. As the current flows through the coil, the field moves with the electrons, exiting one end of the core and returning back around to the opposite end outside the coil, then entering back into the core. This is the flux seen with the filings. In the next figure, the flux on the outside of the coil of both the south and north poles of the magnets flows in unison drawing them together. The flux flows out of the North Pole into the South Pole in unison with no opposition.

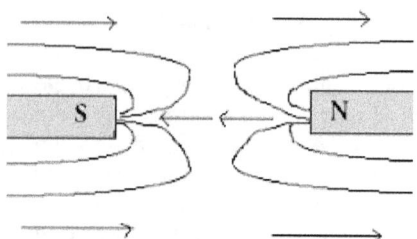

Attraction

The flow of flux out of the north pole of one magnet opposes the flow of flux out of the opposing magnets north pole. Likewise the flux flowing on the perimeter of the coil is in opposite directions. It is these opposing flux flows of the north poles that cause the magnets to repel. The same is true with two south poles of a pair of magnets. In natural magnets an electrical flow within the material itself possibly generates this field.

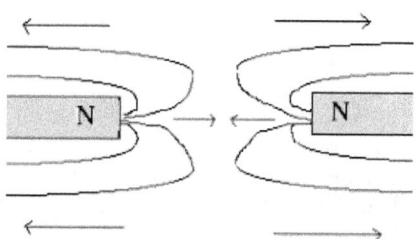

Repelling

Light Propagation Theory

What has been done here in the last few chapters of this book in order to account for the constant speed of light is to compare the propagation of electromagnetic energy to the propagation of sound energy. Since this supposed _medium_ is a gas like substance dispersed throughout space, it moves and is compressed by moving objects. Energy then, as it enters into the _medium_ from an object that is oscillating, moves through the medium at the speed of light in reference to the movement of the medium. Although the velocity of the source emitting the energy may vary, the velocity of light energy through the medium is constant. The velocity of the emitter however is directly proportionate to the wavelength of the energy received at the detector, and may be perceived as red shift or blue shift, an optical Doppler effect very similar to the Doppler effect noticed as a train passes by an observer.

Picture two submarines in the depth of the sea sounding their sonar's in an effort to view obstacles. As Sub 1 is stationary in the water, it pings its surroundings with its 8,000 Hz sonar with the speed of sound in seawater at 3,355 mile per hour producing a wavelength of 0.1875 meters. The sub begins to move and speeds up to 25 knots (28 mph) through the water. As the sonar exits its transmitter into the water, each wave moves forward at the speed of sound in the water. The sub is also moving forward as it sends out the next wave slightly ahead of where it would be if the sub were stationary. As these waves are transmitted through the water, the subs forward velocity compresses the wavelength shorter down to say approximately 0.181 meters, closer together.

A second sub (Sub 2) arrives on the scene, and approaches Sub 1 at 25 knots. As the sound waves passing through the water enter Sub 2's sonar detector, they compress even more by the subs forward velocity through the water to say, 0.175 meters. Now, the speed of sound leaving Sub 1's transmitter was at the speed of sound to that transmitter. Once the sonar wave entered into the water, it transitioned to the speed of sound in the water and was in no way sped up through the water by the subs forward movement. As the sonar waves entered into Sub 2's detector, the sound transitioned from the water into the detector at the speed of sound referenced to the detector. The speed of sound transitioned through three different mediums with the final speed of the sound inside the detector, in no way being influenced by the movement of Sub 1, the water or Sub 2 towards or away from each other.

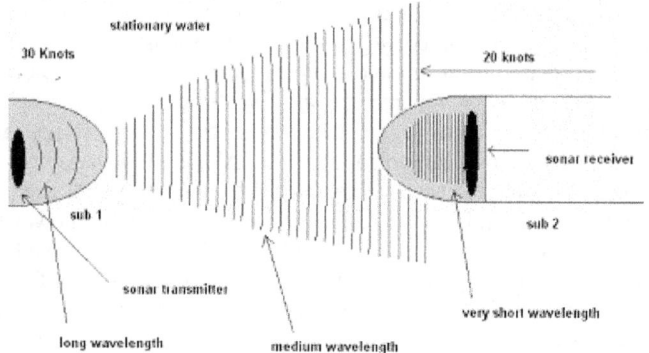

Sound Energy Propagation

In fact, there could be twenty subs all moving at different speeds through the water sending out signals for the other subs to detect. Each signal, as it transitions into the water from the sub transmitting the signal, moves at the speed of sound in the water independent of the subs speed, although the frequencies may vary. Each sub receiving the signal, likewise transitions the sound into its sensors from the water. The speed of sound inside the sensor is the speed of sound in that sensor and is the same for all subs no matter what their speed through the water may be. The frequency received however will vary by the speed of that sub according to the Doppler effect.

The speed of light also is always a component of the medium it is traveling through. As light leaves a star in the great Andromeda Spiral Galaxy, moving toward the earth at about 50 km/s (about 30 mi/s.), transitioning from the atmosphere of the star, out into open space. The light traveling through open space, similar to sound traveling through water is in no way sped up through space by the stars forward movement. As the light enters into the earth's atmosphere, it transitions into the medium moving with the earth and on into the medium in and around the detector. The detector then detects the lights velocity in reference to the medium it is in, not the earth's movement through space. The speed of light transitioned through three different mediums with the final speed of the light in no way influenced by the movement of the star, space or the velocity of the earth.

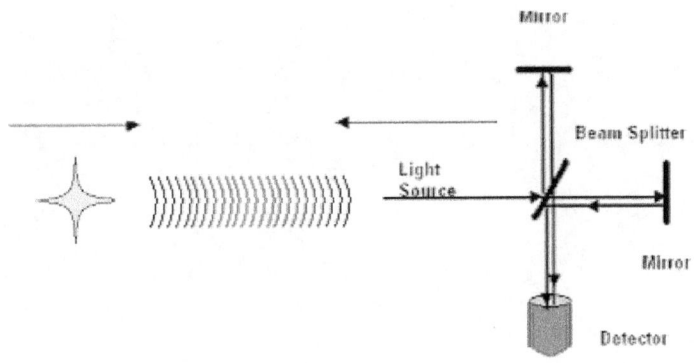

Light Energy Propagation

The speed of light in this context is measured the same no matter what the observers and detectors velocity is on the earth, what velocity the emitter or in this case, the star is, either moving towards or away from the earth and is consistent with Einstein's Theory of Relativity. The velocity of light however, just as with sound, is dependent on the medium that it is traveling through. The perceived wavelength however by the Doppler Effect is dependent and a direct indicator of the velocity of the object the energy left and the velocity of the object detecting it.

Sound vs. Light

It appears then that sound and light are very similar in propagation characteristics. Both seem to emanate in waves and propagate at fixed velocities in reference to the medium that their energy passes through. It also is apparent that the velocity of both sound and light is always the same to the detector no matter what the velocity of the emitter or the detector is in reference to the energy waves. Likewise, moving emitters or detectors generate the common phenomena called the "Doppler Effect."

What then would happen if we took the same mathematical equations developed with the propagation of light theories and applied them to sound wave propagation? Let's look at Einstein's basic formula and simply change the constant "c" to the constant "s" for sound.

$$E = MC^2$$

Energy therefore would be the mass of an object multiplied by the speed of sound squared.

$$20 \text{ kg X } 331\text{m/sec}^2$$

$$\text{Or } 2,191,220 \text{ kg m/sec}$$

Theoretically we could increase the mass by adding energy since the speed of sound is constant; let's go with an energy level of 3,000,000 kg m/sec.

$$\text{Or } 27.4 \text{ kg of mass increase by adding energy}$$

The next formula is very similar:

$$m = \frac{m_0}{\sqrt{1 - \frac{v^2}{c^2}}}$$

With the speed of sound through air at 331 m/sec and a mass of 20 kilograms we calculate mass at half the speed of sound to be:

27390.25/109561=.75; sqrt of .75= .866; 20/.866 = 23.094 kg

Or an increase of 3.1 kg in mass

If we bumped it up to 90% of the speed of sound:

297.9/109561=0.0027; sqrt of .0027 = 0.0521; 20/0.0521 = 383.5 kg

Or a massive increase of 363.5 kg in mass

We could go on and work out these other formulas in the same matter, replacing the "c" with a speed of sound constant "s", with the assumption that nothing can go faster than the speed of sound.

$$t' = \frac{t}{\sqrt{1-\frac{v^2}{c^2}}} \qquad l' = l\sqrt{1-\frac{v^2}{c^2}}$$

$$x' = \frac{x - vt}{\sqrt{1 - \frac{v^2}{c^2}}} \qquad t' = \frac{t - \frac{vx}{c^2}}{\sqrt{1 - \frac{v^2}{c^2}}}$$

$$c' = c_1 + v\left(1 - \frac{1}{n^2}\right)$$

Does this make sense?

Here or There

Zeno of Elea, from the pre Socrates Greek era (late 400's BC), wrote about a race that could not be won, a paradox of sorts. In the story "Achilles and the Tortoise" a tortoise challenges Achilles to a race, as long as Achilles gave him a small head start of only ten meters. The tortoise would then argue that as Achilles would start to catch up, the tortoise would cover a little more distance of which Achilles would also have to also catch up on. This would then go on indefinitely and Achilles could never win.

We can look at this scenario another way; let's start by going from point A to point Z where each step is broken into halfway points. To get to Z from A, you must go through B the first halfway point. Once at B, you must then go through C, the next halfway point. To get to Z you must yet go through D and so on. The points keep getting smaller but you are still not at Z. Theoretically, this can go on for infinity, because you can never get to an end point unless you first go through the next halfway point. Hypothetically then, the smallest divisible points will then be Y and Z. Y being here and Z being there. There is no existence other than these positions, no halfway point. These positions could be the particles of the _medium_. When they move from here to there they move almost instantaneously, not existing unless they are here or there (Wolf Fred, pg.14).

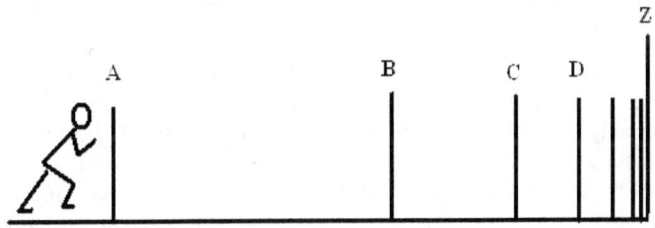

Zeno's Race

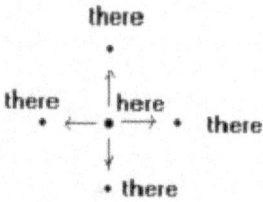

Here and There

In normal everyday life, we can measure say a car and see its location and its state of motion; this also goes for the earth around the sun and a bird in the sky. In the quantum world however, one can only know an objects position or its state of motion (Wolf, pg. 111). Because of the minuteness of this world, even the act of measurement changes the outcome of the measurement. A beam of light for example, being used to measure a particle, will cause the particle being measured to move or change. Therefore, only one measurement such as its velocity can be known, the other desired measurement must be calculated through probability. In order to mathematically calculate these motions and positions, a particle had to be portrayed as a wave function or quantum wave, according to the **Copenhagen Interpretation of Quantum Mechanics** (Wolf, pg. 261).

The act of observation then collapses this wave; probabilities must therefore be used to calculate the position of a particle without directly observing it. Werner Heisenberg pictured the quantum world as a fuzzy, spread out world, a world where specifics could not be known. He stated that the wave nature of particles makes it impossible to know the particles position and momentum. In fact, the more precisely you know a particles position, the less likely you are to know its momentum, only probabilities can be calculated; this is known as the **Heisenberg Uncertainty Principle** (Bord, pg. 388).

Here the very act of observation changes things; even the state of the universe ultimately could be at stake with a change in observation. In fact, according to the Copenhagen Interpretation, until an observation is made, a particles position and state are mere probabilities. Therefore, nothing would exist until it is observed. Einstein would go on to say, "The Heisenberg-Bohr tranquilizing philosophy — or religion? ...Is so delicately contrived that, for the time being, it provides a gentle pillow for the true believer from which he cannot very easily be aroused" (Isaacson, pg. 453). Einstein would ridicule this philosophy as "spiritualistic" and "spooky action at a distance".

Erwin Schrödinger along with Einstein would then look for another way to describe this interaction of particles in the "quantum world". Schrödinger then devised a way to describe particles with his famous wave equation. This equation would explain the probability of a particles position and state when observed (Brown, pg 88).

Einstein did not really like this probability "stuff". He would describe this way of thinking as taking two boxes and placing one ball in one of the boxes; stating that the probability of the ball being in the first box as 50% chance was an incomplete statement to him. The right way to address this was to state that the ball was in the box or the ball was not in the box; that was reality! (Isaacson, pg. 455). Many would argue that the ball was in no means in one of the boxes until they would lift the covers and observed it; the "if a tree falls in the forest" scenario. This was that "religious like" new science that many believed in and Einstein was not so fond of.

Erwin Schrödinger would come up with another analogy similar to Einstein's box. "One can even set up quite ridiculous cases. A cat is penned up in a steel chamber, along with the following diabolical device (which must be secured against direct interference by the cat): in a Geiger counter there is a tiny bit of radioactive substance, so small that perhaps in the course of one hour one of the atoms decays, but also, with equal probability, perhaps none; if it happens, the counter tube discharges and through a relay releases a hammer which shatters a small flask of hydro-cyanic acid. If one has left this entire system to itself for an hour, one would say that the cat still lives if meanwhile no atom has decayed. The first atomic decay would have poisoned it. The psi function for the entire system would express this by having in it the living and the dead cat (pardon the expression) mixed or smeared out in equal parts". It is typical of these cases that an indeterminacy originally restricted to the atomic domain becomes transformed into macroscopic indeterminacy, which can then be *resolved* by direct observation. That prevents us from so naively accepting as valid a "blurred model" for representing reality. In itself it would not embody anything unclear or contradictory. There is a difference between a shaky or out-of-focus photograph and a snapshot of clouds and fog banks" (Isaacson, pg. 456).

Einstein loved it: he would go on to say "Your cat shows that we are in complete agreement… a psi function (wave function – also shown as ψ) that contains the living as well as the dead cat just cannot be taken as a description of a real state of affairs" (Isaacson, pg. 457).

Time and Space

When it became obvious to scientist that the speed of light was not only constant, but also that light was always detected at the same speed no matter what the velocity of the source or detector, scientists became perplexed. Albert Einstein would go on to say that this dilemma had "plunged the conscientiously thoughtful physicist into the greatest of intellectual difficulties" (Einstein, 1952, pg. 18) and came up with a brilliant solution that would remedy all – warp time and space under the effects of gravity and various velocities.

There is another possibility however, that we have been subtly suggesting throughout this book, that may be more reasonable and may work just as well scientifically, another way of altering space and time with the existence of the aether. But didn't the Michelson Morley Interferometer disprove its existence? Doesn't research on Binary stars, gravitational red shift, Mercury's precession and GPS satellite clock anomalies prove that time and space warps with high velocities and strong gravitational forces?

Let's perform another thought experiment where two spacecraft are traveling at two different velocities relative to the same starlight. One spacecraft is traveling towards a star at 150,000 kilometers /second or half the speed of light relative to the star. Another spacecraft is traveling parallel to the first spacecraft at 75,000 kilometers / second or one quarter the speed of light relative to the same star. Both of these spacecraft have observers with test equipment measuring the same light but are traveling at different velocities relative to the star, yet both measure the same light at 300,000 kilometers / second.

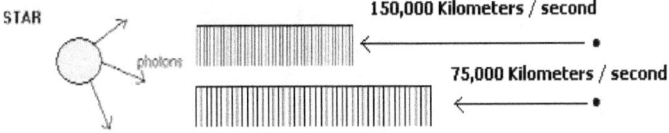

Relative Contraction of Space and Time

What we seem to see here is that the space and time between each spaceship and the star is being compressed and that this compression or contraction of space occurs according to the "Lorentz Contraction" of space relative to each of the spaceships velocity. According to Einstein however, **each spacecraft** would have a different rate of time passage and different measurements on board due to their individual velocities relative to the star; it is not space between the star and the spaceship that is altered, it is the ship itself.

Now, if one spaceship would come between the star and the other spaceship blocking the starlight, suddenly the focus is different. The light the trailing space ship then is viewing is the light reflected off the spaceship in front of them traveling at a slower speed, rather than the light of the star. Did the space ship suddenly shift its shape to adjust to the light reflected off the ship in front of it? If this were true, the ship would have to reshape according to all objects giving off light and moving at different velocities relative to it, compressing for some and expanding for others. Similarly, for binary stars moving in opposite directions to an observer, would both stars stretch and compress at the same time to accommodate for the speed of light to different observers viewing them in different regions of space? Of course not.

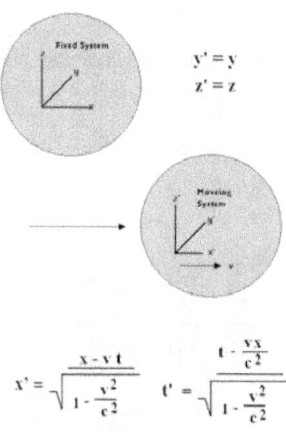

Lorentz Contraction

The velocity of the light detected from an object can only then be relative to the observers themselves; the observer's actions cannot affect the motions or observations of other observers and the space-time around them. Only viewing velocity, time and space in this matter prevents a paradox. This also allows for any velocity to occur without changing space and time; therefore an observer's time and space cannot change with their velocity as previously thought. If time and space is not being warped by the movement of an object, how then is the speed of light the same velocity to all observers, no matter what their velocity or the emitter's velocity?

This brings us back to the submarine illustration. The received sonar signals speed inside the sonar dome of the receiver is at the speed of sound and is constant inside that dome, no matter what the speed of the sub emitting the original signal, or the speed of the sub receiving the signal. The velocity of the water currents between the two subs also has no affect on the sounds velocity inside the sonar dome after it is received. Likewise, the velocity of sound coming down the ear canal of an observer listening to the trains whistle also is unaffected by the trains velocity or the observers own velocity. Sound always has the same velocity; it is only changed by the temperature and pressure of the air it is traveling through. The only thing that can change due to the movement of objects then is the frequency of the detected sound, the Doppler effect as previously reviewed.

The same must be true with light. The velocity of light inside the detector, whether it was Bradley's telescope, Fizeau's and Foucaults instruments, or the Michelson Morley Interferometer will always be the same. This velocity will always be 186,000 miles per second or 300,000 kilometers per second, no matter what the speed of the emitter (train), movement of space between the emitter and receiver (wind) or the movement of the receiver itself (observer). This light moves in waves within some rigid medium and will only vary in velocity depending upon the density of that medium; only the frequency of the light received will vary according to the movement of the various elements, as seen with the Doppler effect as red or blue shift. The two spaceship previously discussed will also have the same process occur. It is not the spaceship that contracts nor is it the space between the star and the ship that contracts, it is the light waves themselves that change as they move from one medium to another. The speed of light inside the detector remains the same, no matter the movement of the ship or the star – just like the submarine.

Time and space here are still affected. The speed of light will vary according to the type of medium it travels through, for example light traveling through glass as compared to light traveling through air. The time it will take light to travel a specific distance will then vary; space also will then vary proportionately with density. Space then also becomes denser close to mass through the cooling process or phase shift as previously discussed and less dense the further away from mass.

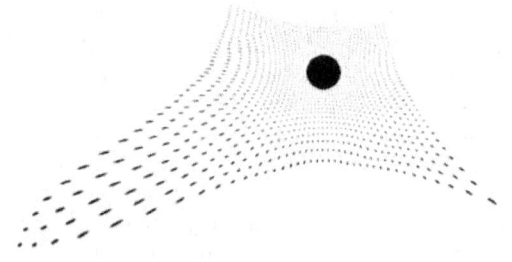

Warped Space

Einstein made an analogy using a bowling ball on a trampoline to describe how space warped with large masses such as our sun. The bowling ball would cause the rubber sheet of the trampoline to sink in the middle, as objects such as billiard balls were rolled onto the rubber sheet; the balls would roll inward toward the bowling ball in a spiral manner (Isaacson, pg. 223). This analogy was Euclidean or two dimensional in description, in real space-time there would be an infinite number of sheets surrounding this ball in three-dimensional space with many layers. In three-dimensional space, it would be spherical sheets much like that of the layers of an onion. Each of the layers of onion would represent a change in gravitational field strength, a change in field strength density if you will. The closer to the bowling ball, the tighter together the sheets become and likewise, the further away the more distance between the rubber sheets or onion layers.

144

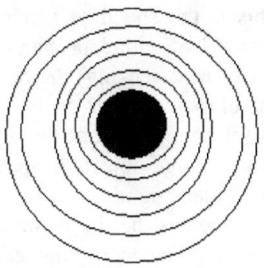

Onion Analogy

What is this field made up of which changes in strength? What would change in density? As we reviewed waves in previous chapters we also asked the question, "What was waving?" Could it be the same material?

In the previous chapter "Here and There," we reviewed that if we were to keep dividing distances in half, which eventually we would come to a point where division was no longer possible; a particle was here and another particle was there, nothing was in between. We can look at these particles as the field that exists in "empty" space, the medium that transfers energy across great distances, and whose condensation creates elementary particles. Could it be the variations in the density of these particles that cause light to bend and the differences in energy levels that create the pressure differential that causes gravity? This could be the onion we are looking for.

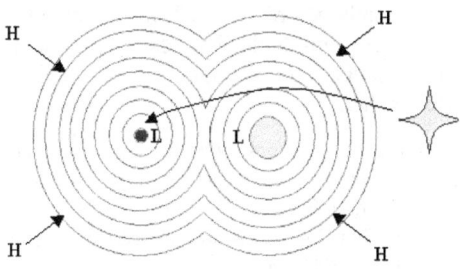

Warping of Space

An example of this is the speed of sound in water. At 100° Celsius (boiling), the speed of sound in water is 1,543 meters per second; however at 0° Celsius (freezing) the speed of sound is 1,402 meters per second. The closer together or the denser these particles are, the longer it takes for energy to pass through them; this is what causes light to bend around large objects in space. The closer to the earth, sun or any other condensing mass such as subatomic particles, the denser the medium becomes that light waves move through; the denser the medium the slower the energy moves.

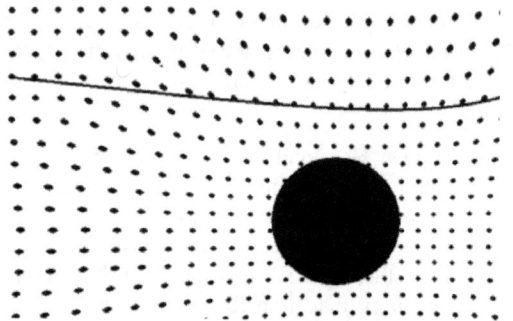

Warped Space Denser Closer to Massive Objects

If mass is merely created out of standing waves within this medium, then the velocity of these waves will also slow down as the medium becomes denser. Motion will be dampened, clocks will tick slower and redshift will be noticeable in light waves emanating out of these objects with large gravitational fields - in fact time will slow down inside of these gravitational fields. This gravitational effect has been clearly noticed in the timing circuits of the GPS satellite system covered earlier.

The larger the mass, the slower the time; likewise the deeper into these onion layers an object moves, the slower the time. Super massive objects such as black holes may indeed slow time down so much in the center that time stops and a singularity similar to the one that existed at the beginning of the "big bang" could exist. The difference here is that this is a miniature "big crunch" or a collapsing of matter, rather the massive expansion of matter during the "big bang." If time goes slower the stronger the gravitational fields, then the reverse should be true outside the gravitational effects of large masses – time should then speed up (Hawking, pg. 90).

The "Theory of Relativity" also states that velocity also changes time. In the section "Blast Off", we reviewed relativity and how that a moving object's velocity is only relative to the position of another object. For example, if a spaceship was far out in the universe with no object to truly reference velocity by, and came across two objects such as asteroids moving away from each other, what would be used to reference the speed of these objects? If they were floating apart at 500 meters per second, was object "A" moving at that speed away from a stationary object "B"? Or was "B" moving away from stationary "A"?

Space Objects in Motion

If "A" was moving faster than stationary "B", time would be slower on "A" but if "A" was stationary and "B" was moving away then "B's" time would be slower. Velocity here is relative and therefore velocity can have no affect on time on the moving objects. There is one catch however, if both objects were stationary relative to each other at one time and one began to move, the object that started to move away had to accelerate. The force of acceleration in the "Equivalence Principle" is the same as gravity – they are indistinguishable. Therefore, it is the acceleration of a moving object that must give it its time change, not velocity itself. Unless a single point of reference for all objects is determined, velocity is simply relative. Only when we have determined that this solid medium that energy waves are moving through exists, then do we have a reference to measure the speed of light to – a stationary point in space.

Velocity, no, rather a change in velocity, well, let's say a constant change in velocity due to acceleration will then slow down time. The longer and the stronger the acceleration forces, the slower time becomes. Therefore, if an astronaut launches out into space and accelerates up to great velocities, time on earth will pass faster than time on the spacecraft. When he arrives back, more time will have passed on earth than in the space ship – he would be younger than if he had never left. If an astronaut has to accelerate leaving the earth, he will likewise also have to decelerate to slow down and return. This deceleration as we saw before, simply changes the direction of the force from aft to forward. The effects of deceleration are identical to acceleration – they therefore are identical. As long as this force exists, time will slow down. Why leave the earth then, why not use a spinning acceleration machine instead to change time? The amount of high forces due to acceleration or gravity may change time but it may be so negligible that it would never be noticed.

We have seen that time may be slowed down due to the effects of gravity or acceleration. Can time be reversed though? Can we go back in time? In his book, A Brief History of Time, Stephen Hawking suggests that even though the Theory of Relativity states that speeds faster than light are impossible, if one could do so, he probably could return before he left (Hawking, pg.163). Simple math however prevents this. No matter how much you multiply a velocity by, you can never get to zero time. For Example, at 100 miles an hour, it will take you one hour to reach the one hundred mile mark. You multiply your velocity by two; you will reach the one hundred mile mark in a half an hour. Multiply by three, one third of an hour and so on. No matter how fast you go, you cannot go faster than time.

Time is not the only thing that changes; space also is changing (which in effect changes time). We studied the possibility of the "Big Bang" how that the universe and all that we know originated out of a singularity, a point where time and space were "infinite". A huge blast then occurred, when the natural balances that contained the singularity broke down, blasting matter and energy out into the far reaches of the universe. Energy, not being able to exist without mass (it is simply the movement of mass) must be associated with some sort of particle, the particle that all mass consists of, the "God

Particle". These particles originally were spread out evenly throughout the universe, but as energy was passed on from one particle to another, this energy was dispersed out and away from the center of the blast. Higher energies to the outside of the blast area forced cooler particles in the center of this area to coalesce and bond together, a phase transition of some sort. This was the foundation blocks of the stars and galaxies that we see today.

As energy slowly is dispersed out and away from the original blast sight, the center continues to cool and contract. Stars and galaxies further away from the center of this blast should therefore have more energy, expanding faster than the cooler areas. Over time, the whole universe should be also cooling uniformly (although areas further away are hotter). Areas further away however, also take longer to view from earth so that objects further out are seen at higher temperatures and at velocities than what they may be today; these areas have since cooled down and slowed down, possibly dramatically.

Edwin Hubble's discovery in 1929 suggested that galactic red shift in light from stars and galaxies increase in proportion to their distance from earth. His studies showed that 38 of 40 galaxies observed are rapidly moving away from each other, an omni-directional galactic expansion where galaxies furthest away are moving the fastest, some at 57,000 kilometers per second or 13 million miles per hour or 2% the speed of light. The problem here is that this is evidently omni-directional, indicating that we are the center of the universe. If we were viewing this expansion out on the edge of the universe, only certain areas focused in one region would have the greatest red shift. Either we are the center of the universe, or this observed redshift is simply an illusion.

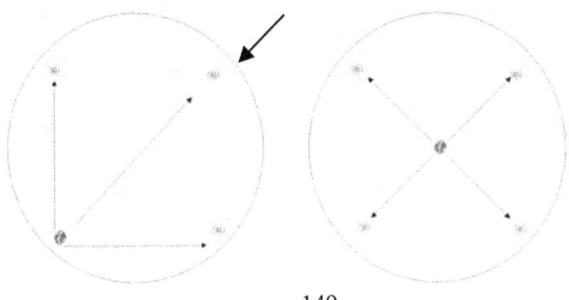

Since it takes time for the universe to cool and also takes time for light to reach us on earth, the changes to the universe will be isotropic or seen uniformly in all directions and be relatively consistent at the same distances. This however suggests a possible "Big Crunch" as energies subside and particles coalesce. Eventually as the medium between all objects condenses and cools, there will be nothing to keep objects apart. Higher energies out side will eventually push all matter together into one place. Here atoms can no longer exist, fusing together into what has been called the **Bose–Einstein Condensate** or BEC's, huge molecules that form at near 0 Kelvin or **absolute** *zero* – cold black holes if you will (Pitaeskii, pg. 1).

Energy however, would have to go somewhere; it cannot just disappear according to the **Principle of the Conservation of Energy**. If space is infinite in all directions, over time all energy would be evenly dispersed omni-directionally and the universe would eventually grow cold and die. Now, if we take the stand that all other matter, including ourselves, were merely energy waves (standing waves) moving through a stationary substance that varied in density, our existence would be limited. The further out in the universe we would go, from the center of the Big Bang, the less dense this substance would be, to the point where matter could no longer exist; much like how sound cannot propagate and exist out in space. Energy cannot move through nothing, therefore energy would have to slowly migrate back into the known universe, reflected back in from the edge of time and space, pressing in on mass in the form of gravity.

If energy however was forced to loop back in at the edge of the universe or loop back around though **black holes** into other universes that constantly feed each other, then it is possible that matter and energy may be able to continue on indefinitely. Many different theories have been developed over the last fifty years trying to show how it all works, from a universe in the shape of a doughnut where time and space loops back in on itself, to the "**String Theory**" of multiple universes. If we think about it, a black hole's singularity draws in matter; the singularity that occurred at the "**big bang**" gave off matter. Can one singularity be the source of the other singularities energy? Is this the entrance to a parallel universe?

Cosmology

Orbital velocities of bodies, in our solar system, normally vary according to Keplerian dynamics; orbital velocities that are proportionate to the distance from the object they orbit. The further away from that object, the slower the orbital velocity of the planet or moon, decreasing this velocity inversely with the square root of the orbital radius from that object, this is due to lower gravitational forces. For example, Mercury takes 88 earth days to orbit the sun while the earth takes 365 days. Mercury's orbital velocity is 48 km/s while the earth's orbital velocity is slower, about 30 km/s in it voyage around the sun. Likewise, Neptune's orbital velocity is only 5.4 km/s and it takes about 164 earth years to orbit the sun.

This, however, is not the case with stars orbiting in a galaxy. Surprisingly, most stars orbit a galaxy at approximately the same speed, in unison, as if tied together; stars on the outer fringes are therefore are orbiting much faster than they should, some at a radial velocity as high as 180 km/s, and theoretically should be slung off into space. As with interactions within our own solar system, orbiting objects are not only attracted to the object they are orbiting; they also have gravitational forces acting on them from all other objects relatively close to them. The earth is, not only attracted to the sun, but all other planets, including our own moon, causing the orbital distance from the sun to vary. Stars likewise, are not only attracted to and orbit around a black hole in the center of the galaxy, they also are gravitationally attracted to all the other stars, planets and gasses moving around the black hole, as if in a soup, moving together. Calculations indicate that there must also be a missing mass that can't be seen in order to account for the necessary increase in gravitational forces needed to prevent these stars from flying off, this is the foundation for the **Dark Matter** theory.

Further calculations had to be formulated to account for additional missing matter, as only 25% of the universes matter could be accounted for. This was the foundation for the **Dark Energy** theory; the energy that helps accelerate stars and galaxies in the far reaches of the universe, faster away from each other, as manifested in Hubble's Law. This law states that velocity, at which galaxies are moving away from the Earth, is proportional to their distance from the earth. The cosmological formula most recently calculated for the universes total mass is 4 percent baryonic matter (composed of atoms and subatomic particles of some sort), 21 percent Dark Matter, and 75 percent Dark Energy. How does this work then? What are Dark Matter and Dark Energy? Let's review the Big Bang.

As discussed previously, the Big Bang theoretically occurred some 10 to 20 billion years ago when something happened within a singularity and thermonuclear energy within the singularity could no longer be contained. A huge amount of energy was released and the smallest of particles blasted out into space at enormous velocities, dissipating their energies out into the blackness of the universe. As this outward energy dissipated, a reversal of the flow or an "implosion" of these particles of mass begins to take place, back toward the center of the blast and towards each other, slowly condensing these smallest of particles into baryonic matter. The outer regions of the universe now contain much of the energy released in this blast as the origin, or center of the blasts begins to cool. At some point in time, the energy blasted out into the universe greatly diminishes and dissipates as the dispersion of energy decreases by the square of the distance it travels away from its source.

The question now being asked is, 'are particles and objects closer to the Big Bangs source moving back towards the origin of the blast, as if sinking into a Black Hole?' If this is the case, could these objects (galaxies, stars, and even us) be "moving away" from the galaxies at the fringes of our universe, rather than those galaxies "moving away" from us? Theoretically, the further away from the origin of the blast an object would be, the weaker the pull on objects towards the center of the Big Bang, giving the illusion of variance in velocity in relationship to distance. As matter condenses and becomes more compact, time also slows down, giving the illusion that galaxies seen at the fringes of our universe (looking back into time) are moving away much faster than they currently are in reality. Could it be that the Universe is no longer expanding and that Hubble's Law is an illusion; an illusion created by an accelerating collapse of "space and time"? The Big Crunch may be underway!

If Hubble's Law were simply an illusion, Dark Energy, the substance that supposedly accelerates galaxies at the edge of our universe, would no longer be required. All that would be needed would be Dark Matter and baryonic matter, materials consisting of atoms and subatomic particles. Dark Matter possibly could be the substance that all baryonic matter is made of, also known as the aether, what all energy may move through.

Hubble's Law – The further away, the faster objects (galaxies) move away from the earth, causing red shift. The earth would have to be at or close to, the center where the Big Bang occurred.

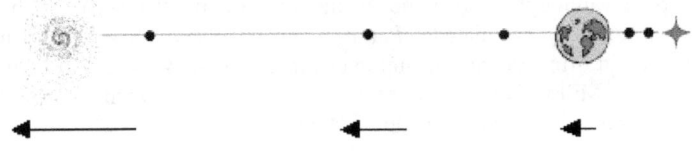

Inverse of Hubble's Law – For mass closer to the start of the Big Bang, the faster mass moves towards the center (as if falling into a Black Hole) creating red shift and causing the illusion that further away objects are moving faster away from us when we are moving away from them.

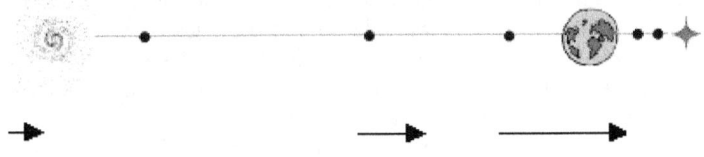

Triangulation has been used for centuries to calculate distances or the height of terrestrial objects, such as trees or mountains. A known length, or height, at a known distance, is compared with an unknown dimension at a greater distance, in order to calculate an approximate measurement; a pencil at arms length compared to a tree for example. One observable variance similar to triangulation with stellar objects is called parallax. Parallax occurs when our viewpoint relative to nearby and distant objects changes, creating an apparent angular change in position of the nearby objects, relative the distant objects. This effect diminishes with distance so that even for the stars nearest to us, the parallax is so small that extremely precise measurements are needed just to detect the effect. Due to the earth's rotation around the sun, the distances to planets such as Jupiter (although it also moves in orbit around the sun) can be approximately calculated. Parallax can be observed as the earth moves in orbit around the sun over time; here stars closest to us in our own Milky Way galaxy seem to wander slightly, when compared to the stars around them that are further away.

However, this process cannot be used to calculate the distance of distant stars and galaxies. Astronomers must use luminosity to calculate the distance to an object, such as a galaxy, by measuring the brightness of the object and applying the inverse square law. The problem is stars vary in size, color and brightness; even galaxies vary in these attributes and are almost impossible to calculate unless you find a standard brightness or candle to go by.

The best standard candle for determining the distance to nearby galaxies is the Cepheid variable star; a star that has a precise level of luminosity and very common in the universe. Some of these stars are even close enough, in our own galaxy, to measure their distance by parallax. With a known distance and a known luminosity, the distances to other Cepheid variable stars in other galaxies can now be calculated. Edwin Hubble used Cepheid variables stars that he recognized in the Andromeda galaxy (our closest neighbor), to approximately calculate its distance from us at 2.5 million light-years. While cataloging the distance to these various galaxies, Hubble discovered that the light signatures from the furthest galaxies away from us had the most red shift; varying proportionately with their distance from us, thereby establishing Hubble's Law.

String Theory

String Theory is one of the latest attempts through very detailed mathematical calculations to explain quantum mechanics using vibrations of one-dimensional strings, which only have length. These strings are the constitutes of all particles and are the foundations of time and space. Various vibrations at different harmonics, such as the strings on a violin that vary in pitch and amplitude, can represent various components of subatomic particles such as quarks. Different combinations of these strings vibrating at different frequencies manifest themselves as electrons, protons and neutrons. Combinations of these strings are called membranes or "**branes**" for short. Different branes may represent different universes and the collision of these branes, like sheets hanging on a clothesline bumping into each other, may be the source of the "Big Bang". Strings can be open or closed; one closed string manifests itself as a graviton, the source of all gravity (Hawking, pg.176).

Open and Closed Strings

Similar to Theodor Kalusa's five dimensional theory proposed in 1919 (Isaacson, pg. 337), M-Theory envisioned in the late 1990's, is one of the latest string type theories that developed 11 different dimensions, 10 spatial and one time: M standing for membrane. Super String Theory is the latest attempt to combine and synchronize the various string theories (five seemingly different theories) into one. It now seems that all of these theories were all explaining the same thing in different ways, all being supposedly correct. All of this was developed with huge mathematical formulas based on the existence of a one-dimensional and two-dimensional world. Equations in the three and four-dimensional worlds became much more complex, explaining the eleven dimensions even more so (Ashtekar, pg. 316).

There is a very strong similarity between what has been suggested in this book and the **String Theory**. We have looked at the possibility that all that exists simply exist as energy waves manifested as **standing waves** moving through some semisolid substance. Solid because as far as we know, transverse waves of any kind can only pass through solids and cannot pass through fluids; light waves are clearly transmitted as transverse waves having polarity. Instead of minute particles oscillating at various frequencies and amplitudes as energy moves through this substance, the String Theory suggests that small one dimensional strings of various lengths both open and closed in structure, vibrate at various frequencies and amplitudes to create the same effect –our world as we know it – a coincidence?

Unified Theory

Until recently, scientific laws have basically obeyed Newton's laws. Scientific discoveries of the 20th century however have uncovered a new science, one that does not seem to obey the same laws. As these different theories evolved, we have slowly developed one sort of law structure for the large and another for the very small. These two theories are somewhat different, as Tim Allen in his book, "I'm Not Really Here" states that the rules, "at a subatomic or quantum level are, well.... different" (Allen, pg. 10).

In his later years, Einstein struggled to make the quantum world and relativity compatible, a struggle to mathematically unify the gravitational and electromagnetic field theories into on **Unified Theory** (Isaacson, pg. 337). "It has been my greatest ambition to resolve the duality of natural laws into unity", Einstein said, "The purpose of my work is to further this simplification and particularly to reduce to one formula the explanation of the field of gravity and of the field of electromagnetism" (Isaacson, pg. 342).

Einstein struggle to his deathbed trying to find the answer for the Unified Theory, writing paper after paper, trying to solve the riddle; but page after page seemed to be flawed, losing the "physical" explanations of his older writings, moving towards "pure mathematics." Scientists of that day began to loose their passion and trust in his writings, and Einstein began to lose his touch, being rejected by many of his colleagues who became skeptical about his work.

Various forces are apparent in the universe, as we have discovered. These are electric fields, magnetic fields, electromagnetic fields, the strong and weak nuclear forces and gravity. All of these forces seem to hold both positive and negative characteristics, either pushing or pulling, except for gravity. Combining these two theories into one became a nightmare! They just didn't work together.

Classical mechanics and Einstein's Theory of Relativity, although there are some things that conflict, they work together much better than the theories of classical mechanics and quantum mechanics; they just did not work together! The **Heisenberg Uncertainty Principle** for example would not work well with classical mechanics. This is where the more precisely you know a particles position, the less likely you are to know its momentum, only probabilities must be calculated.

As we look back at the ideas generated in this book, we see that even through the simplicity of classical mechanics, gravity can be created. Going back to the "Big Bang" and the enormous energies that were released, to the cooling processes that caused mass to coalesce from the building blocks of nature (possibly strings), and the high energies outside of mass that draws particles together.

We also saw how as objects and particles spin, that particles within "empty" space between these particles also may spin (or at least the energy moving within these particles) affecting other objects at a distance. Opposite spins allow objects to move together and not repel while objects of like spins repel each other.

Building this type of universe where laws work together and that both small and large, the regular world and the quantum world move together is not out of the question.

Worm Holes

We have developed a hypothesis that a medium exists in which energy wave's move through; a medium that varies in density according to its "temperature" or "pressure". As this medium cools, it condenses, forming particles that are the foundation blocks of all matter; the closer to matter the medium is, the cooler and denser this medium becomes, in effect changing time and space. We have also hypothesized that matter may also be created by "standing waves" of energy passing through this "elastic-solid" medium. Close to the condensing mass, these energy waves move slower and tend to bend as they pass large bodies of mass, refracting if you will, distorting space and time. The closer to the large mass these standing waves come, the slower they move. This difference in density, creates a pressure differential which in turn causes the gravitational pull; the greater the mass, the greater this pressure differential and gravitational force.

The opposite may be true as these waves move out and away from large masses such as stars and galaxies. Just as light velocity increases as it moves out of the medium of a piece of glass in to air for example, energy moving through "near space" at the velocity of "c" may move much faster far away from large bodies of mass. The velocity of light in space has been assumed to be the constant "c", 186,000 miles per second. However, if space is not empty as we have believed, and light waves move through a medium that changes in density the closer to mass the medium is, and then the speed of light will also change according to its density of this medium it passes through.

If we picture space as a large ocean full of islands, we may also envision an ocean full of currents. As ships on the high seas use these currents today, we may likewise also be able to use the currents in outer space. Areas of space between stars or between galaxies may contain regions where space is almost void of this medium, allowing energy or mass in the form of energy (standing waves) to pass from one region to another at very high speeds – in effect, passing through a natural worm hole.

These wormholes may also possibly be generated mechanically. High beams of energy may be able to temporarily create a void in space, similar to how a lightning bolt superheats and ionizes air momentarily, followed by the sound of thunder. The momentary opening of this tube, mostly void of the medium would allow energy to move through it at speeds much higher than the accepted speed of light of "c", allowing us to "jump" from one region of the universe to another.

This "jump" may also be used for short distances, allowing for transportation from one area on earth to another instantly, "beaming" people or objects from place to place. Energizing or "heating" the medium in close vicinity to an object would decrease the mediums "pressure" on one side of the object. This pressure differential would cause a "gravitational pull" if you will, in a specific direction, moving the object without G forces, as if falling towards the targeted area. If the medium were energized below an object, it would cause the object to rise and be weightless. An object used as a transportation device, could be moved in any direction, and at almost any speed, without the adverse affects of G forces and acceleration /deceleration.

How to energize the medium or aether is the question. Specific wavelengths of electromagnetic energy may be the answer, similar to how water molecules are energized in a microwave oven.

Black Holes

The current Webster's Dictionary definition of a Black Hole is *"a celestial object that has a gravitational field so strong that light cannot escape it and believed to be created due to the collapse of a very massive star"*. The scientific theory of a Black Hole is relatively new. As seen previously, geologist John Michell wrote about the idea over two hundred years ago in his 1783 paper, "Philosophical Transactions of the Royal Society of London". Here, Michell suggested that, if a star were to grow to sufficient mass, at some point, its gravitational force would be great enough to produce so strong of a gravitational field that even light could not escape from it. It wasn't until 1969 that the term was coined when scientist John Wheeler used it in a speech, and stuck (Hawking, p. 84).

Soon, new ideas began to surface and Black Holes started to become more scientifically complex. The core of the collapsed star now had a "singularity" similar origin of the Big Bang, and the outer shell of the Black Hole was now called the Event Horizon. New studies of galaxies showed high velocity stars spinning near the center of these galaxies, and x-ray emitting gasses traveling close to the speed of light, seemingly proving that Black Hole's existence (Bartusiak, p.210).

In his book, 'Hawking on the Big Bang and Black Holes', Stephen Hawking explains empty space as not truly empty, but full of virtual particle pairs consisting of both positive and negative energy. Hawking goes on to explain that as these particles begin to be pulled into the Black Hole's Event Horizon, the pairs split up; the positive particle being ejected out from the Black Hole as "thermal emission" while the negative particle "tunnels" uneventfully into the Black Hole (Hawking, Hawking on the Big Bang, p.88).

Although Hawking does not subscribe to the "Aether Theory", he seems to indicate that "empty space" is not empty after all and that "local energy density" varies with varying amounts of these particles from place to place (p.87). This strongly parallels the hypothesis that is being promoted by this book, which all matter consists of condensates of some material, which energy cannot exist on its own, and is simply manifested by the movement of this material.

Hawking also suggests that the positive energy particle is being ejected away from Black Hole as "thermal emission" or energy radiation of some sort. Could it be that Black Holes are simply made up of the predicted, Bose-Einstein material, which is created by mass so super cooled, and that it collapses into its self as one solid material? Super cooled due to the fact that matter has all the energy "wrung out of it" as radiant energy, before it falls into the Black Hole?

Theoretically, as mass approaches a Black Hole, energy literally is squeezed out of it; evidenced by the x-rays and the brightness of the center of galaxies such as MCG6-30-15. Here, it would not be the attraction of the super massive Black Hole that pulls in all material, but the huge differential in pressure between energetic particles in the universe along with cold condensed particles in the Black Hole. The universe pushes in, in order to create gravity, not the mass of the Black Hole that pulls in on the matter surrounding it. As energy is squeezed out of these particles, the particles move slower and slower until all energy subsides and unifies itself with the huge Bose-Einstein mass of the Black Hole.

The Black Hole is therefore "black" not because of its intense gravity that prevents light from escaping, but simply because it has no energy at all; as all energy is forced out of mass before it enters the Black Hole. Again, this suggests that the universe is cooling and contracting as energy bleeds away from its core. The expanding universe may again, simply be an illusion, and the Big Crunch is under way.

Review of Theory

Einstein and the other scientist at the turn of the 19th century were baffled by the fact that the speed of light was always the same velocity to an observer, no matter what the velocity of the observer or the velocity of the object emitting the light. Lorentz had shown that if an object was moving, that it contracted in the direction of its motion, increasing the distance light had to travel, therefore allowing the "perceived "speed of light to remain the same. Motion was warping space and time. This led Einstein to develop his theories of Relativity.

The Theory of Special Relativity explains that motion is relative; Einstein showed this with the train and embankment thought experiment. Here two bolts of lightning strike simultaneously in front and behind a moving train. There is one observer midpoint between the lightning strikes on the moving train and one observer midpoint the lightning strikes on the embankment beside the train. Although the lightning strikes occurred at exactly the same time to the observer on the ground, the observer on the moving train saw the light from the front of the train first because of his forward motion. By the time the light meets this observer, he has moved towards the lighting in front of the train and away from the lightning behind the train.

In this text, we have reviewed a similar thought experiment with a train blowing its whistles simultaneously from the engine and caboose. We saw the different times the whistle was heard by the two observers and also saw the Doppler Effect by the movement of the train. Here, movement or motion caused a different reality for both observers.

The Theory of General Relativity went on to show that mass also has an effect on time and space. Mass causes space to bend around it, which in turn causes mass to attract mass, the effects of gravity. This also is what causes light waves to bend and slow down, manifesting itself in gravitational lensing and red shift. The effects of mass also cause clocks to change their tempo, in effect changing time. Therefore both mass and velocity have a direct influence on time and space. The greater the mass, the more space and time are warped and the greater the velocity, the greater time and space are warped.

Einstein also showed that gravity and acceleration are equivalent with his Equivalence Principle. This was shown with a thought experiment where an observer was floating in space in a large room with no gravity. A force was applied to the room that caused it to accelerate at a constant rate, forcing the observer onto the floor. The observer, not being able to see outside, assumed that this force was gravity. This asserted that gravity and acceleration is essentially the same thing.

Another thought experiment was performed in this text in Blast Off. Here velocity was studied comparing the speed of one rocket stage to another in outer space, showing that velocity was relative to a reference point – one of the other stages. This point was also shown with the baseball being thrown from a moving vehicle earlier in this text. The point here given was to show that if speed was relative, the effects of speed would also have to be relative. This means that since velocity changes time, that this time change would be relative to different observers at different relative speeds.

As mass increases at speeds close to the speed of light, this mass change must also be relative to different observers. An object would therefore be getting larger to one observer than to another observer at a different relative velocity; how can this be? Just as in Einstein's train, this can only be a **perceived** difference and not reality. The only way that we can get to the point where these effects of velocity are real, is to create a stationary point in space in which all object's velocities are referenced to.

Let's Talk About Relativity

Let's look at this again. We all know the scenario where two cars are traveling toward each other; each car is traveling at 50 mph. That is 50 mph relative to the ground. Relative to each other they are traveling at 100 mph. As they pass, one passenger will throw a baseball to a passenger in the other car at 20 mph, relative to the car it is thrown from. Relative to the passenger catching the ball, say he had a radar detector aimed at the ball, the ball will be traveling at 120 mph. If an observer standing at the side of the road aimed a radar detector at the flying ball, it would read 70 mph (50 + 20). Of course here on earth we "usually" use the earth as the reference for speed. An aircraft on the other hand must use airspeed because the lift of the aircraft depends on it. In space, outside our solar system, what is our reference for velocity? Let's put a satellite up in space and launch a rocket from it at 80,000 miles per second (mps). Let us now launch a smaller rocket from this rocket at 80,000 mps. Then another rocket from this rocket is launched at 80,000 mps. The third rocket should be traveling at 240,000 miles per second in reference to satellite from which the original rocket was launched.

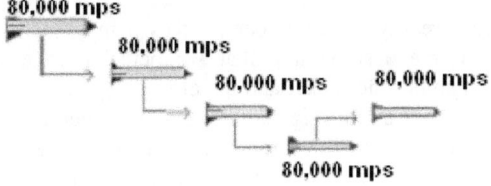

Rockets

Now what is the speed of light? 186,000 mps! What is it relative to? Electromagnetic energy (light, radio waves, x-rays) is always at the speed of light when observed, no matter what the position or velocity of the observer or receiver. There is no other reference point. The movement of any object is always measured by its change of position over a given period of time in reference to something; a change that may be in reference to its original position in space. Any object in space can be considered to be stationary unless it is changing in acceleration (speeding up or slowing down). In the illustration above, all objects are moving slower than the speed of light in reference to each other except, the first rocket launch sight (the satellite) relative to the last rocket. Neither the satellite nor the last rocket are accelerating or decelerating. The first may consider itself stationary or the last may consider itself stationary. The first rocket may be moving away from the last at a speed above the speed of light or the last rocket may be moving away from the first at a speed above the speed of light, which is it?

Einstein stated that an object's mass increases with its velocity. But what is its velocity relative to? He must be referring to stationary space! What is stationary space? How fast is our system moving relative to stationary space? If an object is moving at a high-speed relative stationary space, its mass increases. Does that mean that if an object slows down to zero velocity relative to stationary space, it will be non-existent and have no mass?

If there is no state as stationary space, and the last rocket moves away from the first at the speed of light, then the last rockets mass increases. If the last rockets mass increases, the first rockets mass becomes smaller relative to the last; or is the first moving away from last at the speed of light? If this is the case, then the first rocket should be getting larger, not smaller, in relation to the last rocket.

I'm afraid reality is not as complex as it may seem. Let's be careful not to make something that we perceive to exist, to be absolute. Remember the rocket and the clock? As the rocket moving away from us sends a radio signal of its clock, we perceive the clock to be ticking slower. As it returns, it is perceived to be faster. But reality is different. The clock never changed its rate of ticking. It never changed time!

This *medium*, the "aether", must exist. The whole electro-magnetic spectrum is dependent on it. This *medium* is mass and relatively stationary in undisturbed space. The smallest of all particles, it may consist of all three energy states, positive, negative and neutral. The neutral state may simply be a combination of both positive and negative states. The *medium* particles maintain an equidistant spacing from each other, depending on the density of the surrounding space. The *medium* is contained in both mass and empty space. Within mass and close to strong gravitational fields, the *medium* is much denser and will effect the transmission of electromagnetic waves through it. The velocity of electromagnetic propagation depends on the density of the *medium*. In materials that have high density, the *medium* is denser, such as in glass and plastics, and the velocity of the energy through them is slightly slower. The *medium* is an elastic solid, unlike the assumptions of Michelson and Morley, and is compressed by moving objects, gravitational, electric and magnetic fields. It is in fact the change in density and pressure of this medium that generates these fields. Outside of any movement and external fields, such as magnetic of gravitational fields, the *medium* falls to a stationary and neutral "pressure". In front of a moving object, the *medium* compresses and therefore increases in "pressure". In the wake, it falls to negative pressure and expands; the "pressure", being the ratio of the spacing of its particles in reference to stationary, neutral space. All electromagnetic energy is transferred through the *medium* from an emitter to a receiver. The number of particles in a given space determines the velocity of the electromagnetic waves propagation. The denser the *medium*, the less elastic the *medium* becomes and slower the electromagnetic propagation.

Stephen Hawking, in his book, A Brief History of Time, suggests that there may be a form of matter that is uniformly distributed, yet still undetected that could raise the average density of the universe (Hawking, pg. 48). Could this be the medium that bends light from stars and slows down radar signals bounced off planets? Could this be the dust that causes Mercury to precess more than expected in its orbit close to the sun? Is this the force that changes the time in the GPS satellites or the extra mass contained in the galaxies that prevents the outer stars from flying out into the open universe?

Conclusion

The **first assumption** we will make is that energy is the movement of mass. Scientifically speaking, energy is a variable quantity that is a characteristic of an object (mass). There are several ways to describe energy, kinetic, potential, gravitational, electromagnetic, chemical and nuclear. Energy can be transformed from one form of energy to another such as light into heat, but cannot be created nor destroyed. Action at a distance, such as the transfer of energy as light or gravity must therefore be transferred through the movement of mass; this may involve the photon and the graviton as taught in modern physics.

Light moves in waves and clearly generates the Doppler effect when objects that give off energy move. Waves move through some sort of medium that is relatively stationary in space; an example is waves in water and sound through air or any other object such as steel. Here the **second assumption** is made that light waves must move through <u>something</u> that is relatively stationary in space.

Light waves are transmitted in both longitudinal and transverse waves. Longitudinal waves can move through any solid or fluid, however, transverse waves can only move through solids. The **third assumption** therefore states that the substance that light waves moves through must be a solid – generating the problem of how planets and stars can move in rigid space.

If space is rigid, then all movement must consist of energy waves passing from one particle of this semi rigid substance to another. Standing waves manifest themselves normally as interference patterns generated in two or more waves of similar harmonics, interacting with each other in a stationary medium. Our **fourth assumption** is that matter may be simply manifested as energy waves moving through a solid medium. A lenticular cloud forming over a mountain ridge may seem to be created of stationary particles. This cloud however is formed by moving moist air moving over a mountain top, condensing its moisture in the cold winds at high altitude as the air drops below the dew point. As the air moves down the downwind slope it warms up and evaporates. This constant movement of air creates a "standing wave", a cloud that seems to stand still, however generated by fast moving condensing then evaporating water particles.

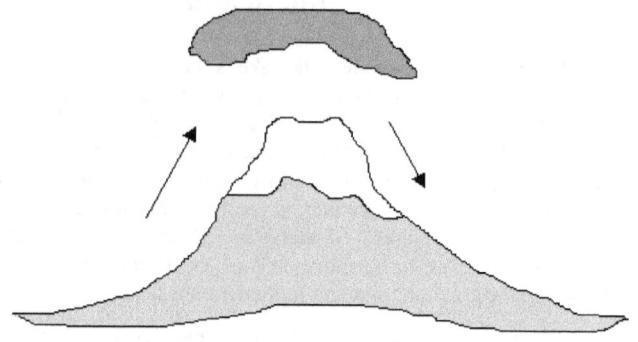

Lenticular Cloud

Another form of a standing wave can be seen as vapor trails over the wing of an airplane especially on a moist or rainy day. Here, similar to the lenticular cloud, moving air condenses in the low pressure region above the wing. Although this condensation seems to be stationary to the passengers in the aircraft, the air moving through this area may be moving at speeds over 500 miles per hour. Moisture in the moving air condenses momentarily in the low pressure area and then disappears as it moves away from the wing creating a standing wave.

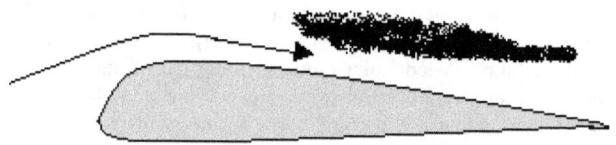

Low Pressure Condensation

Similarly, photons and electrons may simply be the manifestations of confined energy. Energy moving as waves at the quantum levels must be transferred in packets of Plank's Constant and cannot be divided. Energies at these quanta will then move in a straight line or possibly in a confined, circular orbit – as if trapped; it cannot be dispersed. This trapped or confined energy then manifests itself as a moving particle in a "stationary" medium.

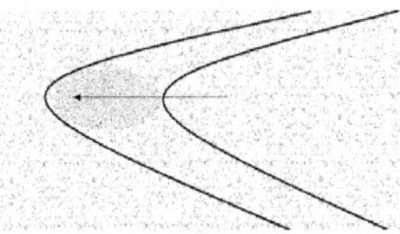

Photon Moving as a Particle in a Solid Medium

Energy of a photon may move from particle to particle in a stationary medium, yet confined to a specific size and area due to Planks Law. The particle moves in a straight line at the speed of light until absorbed by some other substance, normally an atom of some sort. A photon would therefore manifest itself as a moving particle even though it would simply be energy moving though a stationary solid substance.

An electron would likewise move through this medium as the photon does, however, it would only exist at specific regions known as shells around the atom. As it would possibly gain angular momentum, due to absorption of energy in the form of a photon, it will reappear at another shell further out. Energy absorbed above a specific level will eject this extra energy as a photon away from the atoms nucleus. An electron would therefore manifest itself as a moving particle even though it would simply be confined energy moving though a stationary solid medium around a condensed particle or group of particles – the atom.

It is possible however, that Keplerian dynamics and angular momentum do not pay a role as we may think with these particles. Since these particles may simply be the manifestation of energy, there is no true net movement of mass, just waves of energy moving through a solid substance. The velocity of the electron wave may therefore increase in speed the further away it moves from the center of mass or atom; as the medium becomes less dense and the closest electrons to the atom may move the slowest in the densest medium.

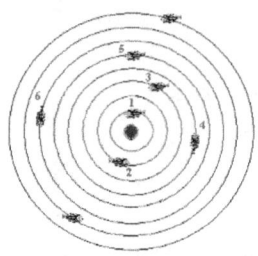

Electrons as Waves Moving Around a Nucleus

Without this stationary substance that light waves would move through, there would be no reference for the speed of light. Photons as moving particles would move relative to the object ejecting them or absorbing them, and light could not move at a constant velocity. Dual stars orbiting each other and moving in opposite directions as seen by Willem de Sitter, would give off light at different speeds to different observers. The mere fact that light generates Doppler shift is proof that light moves as a wave through a medium. Light or any other form of electromagnetic energy would therefore always move at a specific speed relative to the observer, similar to the speed of sound moving in the submarines sonar detectors dome or the speed of sound in the ear canal of someone hearing a trains whistle.

The **fifth assumption** will then state that other forms of mass would form from the cooling and coalescing of particles of this substance that is dispersed throughout the universe. This substance would contain hot and cold areas, cold areas having more densely spaced particles in this substance with a lower "pressure". This pressure differential of low pressure close to particles and high pressure away from particles will cause particles to move together –

172

the generation of gravity; the more the mass, the higher the differential pressures pushing mass together.

We can visualize this process by a simple thought experiment. Imagine if you will a room separated in the middle by a thin plastic membrane, sealed off so that air cannot pass from one side of the room to the other. The room will be quite warm, say 38^0C or 100^0F and very humid. We then, using a heat exchanger with a refrigerant of some sort, consisting of a condenser and evaporator, transfer heat energy from one side of the room to the other with no net air movement; both sides air tight and insulated. Over time, one side of the room will become cooler and cooler as the other side of the room becomes hotter and hotter. Soon the cold side of the room begins to fog up as water vapor starts to condense in the air and droplets of water begin to form on cooler surfaces; it doesn't take long until frost begins to form on the coldest objects in the room. Now, take notice of the thin plastic membrane, how it now bulges into the cool area of the room. Here, the hot air, having higher pressure, pushes in on the cold area, ballooning if you will, the membrane into the cold room; the hot region, pushing in on the cold region.

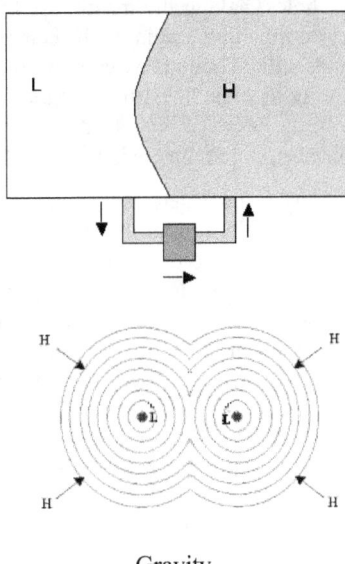

Gravity

173

Back to the real world, energy such as light waves moving through these denser regions, would move slower and refract, bending the light, similar to how light moving from air into water in a glass, makes a spoon seem to bend. This creates a **sixth assumption** that the speed of light varies according to the density of the medium it passes through, higher densities being close to mass.

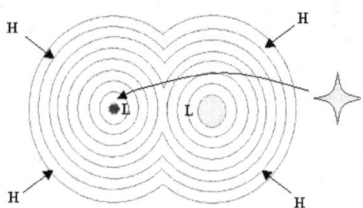

Starlight Bent by the Suns Gravity

The **seventh assumption** then states that if light moves slower close to large masses and that time itself will move slower. The opposite would then be true that the lack of gravity would in fact speed up time. Black holes, having the most mass therefore would have the largest effect on time to the point at the center; a singularity exists where time stands still. Time may vary from region to region, each area of space having its own "local time"; time however, could never be reversed, only stopped or sped up. Could these singularities be the portals to other dimensions?

Just as with sound in air or water, energy moving through this medium in space will move at a constant velocity depending on the "density" and "temperature" of the medium. Even though nothing that consists of energy waves could move faster than the speed of light, the speed of light will still vary from one region of the universe to another depending on the density of this medium. This then would suggest the **eighth assumption** that wormholes do exist. Wormholes would be areas in the universe almost void of this medium. Energy moving in this area could move at velocities thousands of times faster that the speed of light in "near space" or space close to massive objects such as our sun. If mass merely consists of energy waves moving through this medium, then mass

174

could vary well be able to move through these areas at very high speeds, allowing one to "pop out" on the other side of the universe.

If this medium is dispersed throughout the universe in all directions and was created by the "Big Bang", it then should become less dense, the further away from the center of the source of the energy, to the point where it no longer exists. At this point energy can no longer propagate (similar to how sound waves cannot propagate out into space) and matter and time can no longer exist – the edge of the universe, the **ninth assumption**.

The **tenth assumption** is that this universe is slowly collapsing. Edwin Hubble's discovery in 1929 suggested that galactic red shift in light from stars and galaxies increases in proportion to their distance from earth. His studies showed that 38 of 40 galaxies observed are rapidly moving away from each other, an omni-directional galactic expansion. Galaxies furthest away are moving the fastest – some at 57,000 kilometers per second or 13 million miles per hour – 2% the speed of light. The problem here is that this is evidently omni-directional – indicating that we are the center of the universe. If we were viewing this expansion out on the edge, only certain areas focused in one region would have this red shift. It must be that either we are the "center of the universe" or this observed redshift, therefore, is simply an illusion; an illusion due to the shrinking of the universe around us as it cools, loosing energy out into the vastness of the universe.

Since it takes time for the universe to cool and also takes time for light to reach us on earth, the changes to the universe will be seen uniformly in all directions and be relatively consistent at various distances. This however suggests a possible "Big Crunch" as energies subside and particles coalesce. Eventually as the medium between all objects condenses and cools, there will be nothing to keep objects apart; this expansion medium will no longer exist. Higher energies out side will eventually push all matter together into one place. Here atoms can no longer exist, fusing together into what has been called the **Bose–Einstein Condensate** or BEC's, huge molecules that form at near 0 Kelvin or absolute zero and possibly what Black Holes are made of.

176

Summary

In summary, our **first** hypothesis is that energy cannot exist without mass, much like velocity cannot exist without the bullet. There can be no such thing as pure energy; mass cannot become energy and energy cannot become mass. Energy is simply manifested as the movement of mass in some way or form.

Secondly, we hypothesize that light waves must move through something, much like waves move through water or air; something must be waving, a medium or field, sometimes called the aether.

The **third** hypothesis is that, whatever light or electromagnetic energy is moving through (the medium or aether as it might be called) must be considered to be an elastic-solid. This is due to the fact that transverse waves, waves that are polarized, can only move through solids and cannot move through fluids.

The **fourth** hypothesis is that mass consists of standing waves within this elastic-solid medium. This is why the planets, stars and galaxies are not rigidly in place and move freely within this medium. Picture the universe, if you will, as a three-dimensional television screen, where individual pixels close up, do not move. Simply speaking, the energy that is changing the pixels color and brightness, changes in amplitude and moves from pixel to pixel, giving the illusion of movement while the pixels remain solidly in place.

The **fifth** hypothesis is that particles of the subatomic world are this illusion; energy moving through this medium created by the cooling and coalescing of these particles or confined pressure waves moving through it. This is why the Michelson Morley experiment failed; mass is the medium, manifested by energy moving through it. Pressure differentials in this medium create the illusive condition called gravity. Here, higher energies away from mass press in on the lower energies of colder, condensed mass (including the fiery stars), this causes mass to attract mass.

Our **sixth** hypothesis states that closer the medium is to the coalescing mass, the denser the medium becomes, causing light passing through this area to change velocity. Since light moves in waves, refraction occurs, causing light wave to bend as they pass close to large masses.

Our **seventh** hypothesis is that time will vary with gravitation, creating what is called "local time". Time, is the change in position of energy moving through the medium, frame by frame. Energy, however, moves slower closer to mass due to the mediums increased density. Time therefore will move slower the closer to mass an object sensitive to time comes. Time also will vary during acceleration and deceleration due to the "principle of equivalence" where gravity and acceleration are indistinguishable; it is not velocity that changes time, it is the acceleration or deceleration to that velocity of the object that changes "local time".

Our **eighth** hypothesis is that wormholes do exist. In the far reaches of the universe where the mediums density is almost void, energy may move at velocities much higher than our calculated speed of light. Since mass is the movement of energy through this medium, mass in this region may simply exist, then not exist, and then exist again, "jumping" as it were, far from where it came from. This jump is almost identical to the jump the electron encounters as it jumps from one shell to another, existing, then not existing, then existing again.

This brings us to our **ninth** hypothesis that at some point far away from the center of the "Big Bang" that mass can no longer exist, a place where space is completely void of this medium. Small wormholes may also be created here at home, allowing one to "beam" from one place to another. This could be done by "energizing" or heating this medium in close vicinity to a transport vehicle. Energizing the medium would create pressure differentials in the medium similar to gravitational forces explained earlier. Since the medium that is being energized or heated, consists of so little mass, very little energy may possible be necessary to create this force. It may simply be a matter of frequency, similar to a microwave oven heating your diner. This technology may allow the craft to hover or move at extreme velocities without the negative effects of g forces on its occupants, as if "falling" toward its intended destination, possibly falling into the sky.

178

Our **tenth** hypothesis is that the universe is slowly collapsing. Redshift seen from galaxies furthest away from us suggests that rather than having an expanding universe, the universe is slowly collapsing. As energy in our universes core is radiated out into the outer reaches of space, energy is being lost and our core is cooling and shrinking. As we look back into time, an illusion is created that suggests that these galaxies are racing away from us omni directionally; in fact, it could be that we are racing away from them, as if falling into a "black hole". Black holes are all around us, the centers of probably all galaxies. Could it be that black holes are regions where the universe is gone completely cold? Where energy in mass has been completely "wrung out" and the "**Bose Einstein Condensate**" is created? Is this the beginning of the "Big Crunch"?

Appendix

I cannot but regard the aether, which can be the seat of an electromagnetic field with its energy and its vibrations, as endowed with a certain degree of substantiality, however different it may be from all ordinary matter.

(Lorentz, 1906)

(Borel, Armand, pg. 690)

Indeed one of the most important of our fundamental aether not only occupies all space between molecules, atoms, or electrons, but that it pervades all these particles. We shall add the hypothesis that, though the particles may move, the aether always remains at rest.

(Lorentz, 1906)

(Lorentz, H.A, pg. 11)

Lorentz proclaimed the very radical thesis, which had never before been asserted with such definiteness: the aether at rest in absolute space.

In principle this identifies the aether with absolute space. Absolute space is no vacuum, but something with definite properties whose state is described with the help of two directed quantities, the electrical field E and the magnetic field H, and as such, it is called the ether.

(Born, 1924)

(Born, Max, pg. 204)

Index

47, 49, 58, 61, 65, 67, 68, 72, 73, 76, 82, 91, 95, 98, 105, 106, 107, 131, 132, 133, 141, 142, 143, 144, 147, 149, 159, 160, 161, 163, 164, 165, 166, 171, 172, 174, 175, 178

Spin, 35, 56, 65, 78, 119, 120, 121, 126, 158

Standing Waves, 60, 146, 150, 156, 159, 169, 177

String Theory, 150, 155, 156

Strong Nuclear Force, 109, 112

Subatomic, 35, 38, 53, 58, 64, 78, 79, 106, 118, 119, 121, 146, 152, 153, 155, 157, 177

Submarine, 107, 143

Telescope, 2, 143

Theory Of Relativity, 3, 5, 14, 23, 104, 133, 147, 148, 157

Thomas, 1, 67

Transverse Waves, 1, 76, 156, 169, 177

Twin Stars, 2

Undulation Theory, 1

Unified Theory, 157

Velocity, 2, 3, 5, 7, 8, 9, 13, 14, 15, 16, 18, 19, 23, 26, 28, 30, 33, 34, 36, 37, 38, 39, 44, 51, 57, 62, 63, 65, 68, 80, 81, 84, 85, 86, 88, 89, 90, 91, 97, 103, 104, 105, 107, 108, 109, 121, 122, 126, 131, 132, 133, 135, 138, 141, 142, 143, 146, 147, 148, 151,152, 159, 161, 163, 164, 165, 166, 172, 174, 177, 178

Venus, 3, 123

Warp, 5, 23, 26, 141

Wave Theory, 1

Wavelength, 36, 62, 63, 67, 77, 81, 82, 85, 87, 88, 89, 93, 95, 97, 98, 103, 107, 131, 133

Waves, 1, 3, 32, 58, 59, 60, 62, 64, 65, 67, 68, 69, 70, 71, 72, 73, 74, 75, 77, 79, 82, 87, 97, 99, 100, 101, 103, 105, 106, 107, 108, 109, 112, 113, 121, 123, 124, 131, 135, 143, 145, 146, 147, 150, 156, 159, 163, 165, 167, 169, 171, 172, 174, 175, 177, 178

Weak Nuclear Force, 112, 157

Zeno Of Elea, 137

References

Allen, Tim (1996) I'm Not Really Here, Hyperion, 114 Fifth Avenue, New York, NY

Arago, François (1858) Oeuvres Complètes, 7th Edition, Volume 4, Paris

Aristotle (350 B.C.E) De Caelo, Book I, Chapter 2

Ashtekar, Abhay (2005) One Hundred Years of Relativity, World Scientific Publishing, 5 Toh Tuck LinkSingapore

Asimov, Isaac (1991) Atom, Journey Across the Subatomic Cosmos, Truman Talley Books, First Plume Printing, New York, NY

Barbour, Julian B. (2001) End of Time: The Next Revolution in Physics, Oxford University Press, 198 Madison Avenue New York, NY

Bartusiak, Marcia (2000) Einstein's Unfinished Symphony, Listening to the Sounds of Spac^{2}e -Time, Burkley Publishing Group, New York, NY

Bless, Robert C. (1996) Discovering the Cosmos, University Science Books, 55D Five Gates Rd, Sausalito CA

Bord, Donald J. (2000) Inquiry Into Physics, Brooks / Cole Fourth Edition.

Borel, Armand (2001) Oeuvres, Collected Papers, Vol. IV, Springer Scientific and Business Media, 1150 65th Street, Emeryville, CA

Born, Max, (1962) Einstein's Theory of Relativity, Dover Publications Inc. New York, NY

Brown, Julian, (2000) Minds, Machines, and the Multiverse, The Quest for the Quantum Computer, Simon and Schuster, New York, N.Y.

Collins, Harry (1998) The Golem: What You Should Know about Science, Cambridge University Press, Cambridge CB2 2RU, UK

Dekel, Avishai (1999) Formation of Structure in the Universe, Cambridge University Press, Cambridge CB2 2RU, UK

Dennis, Mark F. (2009) The Speed of Light, Reviewing the history of "c", Golden Iris Publications

Ditchburn, R.W. (1991) Light, Dover Publications, Inc., 31 E.2nd Street, Mineola, NY

Einstein, Albert, (1905) On the Electrodynamics of Moving Bodies, June 30, 1905

Einstein, Albert, (1952) Relativity by Einstein, Special and General Theory, Tess Press, Fifth Edition

Evans, Alvis, J. (1985) Basic Electronics Technology, Tandy Corporation, Ft. Worth, Texas

Fresnel, A. (1818) Lettre d'Augustin Fresnel à François Arago Annales de chimie et de physique. 9, pg. 57–66

Galileo, Galilei (1632) Dialogue Concerning the Two Chief World Systems, Second Edition, University of California Press, Berkeley, CA

Glendenning, Norman K. (2007) Our Place in the Universe, World Scientific Publishing, 27 Warren Street, Hackensack, NJ

Hawking, Stephen (1988) A Brief History of Time, Bantam Books, Random House Inc. New York, N.Y.

Hawking, Stephen (1993) Hawking on the Big Bang and Black Holes, World Scientific Publishing, P.O. Box 128, Farrer Rd., Singapore

Hobson, Michael P. (2006) General relativity: An Introduction for Physicists, Cambridge University Press, Cambridge University Press, Cambridge CB2 2RU, UK

Isaacson, Walter (2007) Einstein, His Life and Universe, Simon and Schuster Paperbacks, New York, NY

Kafatos, Menas (1990) The Conscious Universe, Springer-Verlag, New York, NY

Lorentz, H.A, (1909) The Theory of Electrons, B.G. Teubner, G. E. Stechert, Booksellers, New York, 129- 133 West 20th St. New York, NY

Newton, Sir Isaac, (1730) Opticks, a Treatise of the Reflections, Refractions, Inflections and Colors of Light, Fourth Edition.

Pitaeskii, Lev (2003) Bose-Einstein Condensation, Oxford University Press, USA, 198 Madison Avenue New York, NY

Poincaré, H. (1904) The Present and the Future of Mathematical Physics, Address Delivered to the International Congress of Arts and Science, St Louis.

Polhinghorne, J.C. (1984) The Quantum World, Princeton Science Library, Princeton University Press, Princeton NJ

Psillos, Stathis (1999) Scientific Realism, Routledge, 29 W. 35th Street, New York, NY

Sokolov G., (1999) The Theory of Relativity and Physical Reality

Tippler, Paul (2003) Physics for Scientists and Engineers, Standard Version, W. H. Freeman Company, 41 Madison Avenue, New York, NY

Veatch, Henry (1978) Electric Circuit Action, Science Research Associates, Chicago Ill.

Van der Kamp, Walter (1988) De Labore Solis, Airy's Failure Reconsidered, Anchor Book & Printing Centre, Surrey, B.C., Canada

Wolf, Fred, (1989), Taking the Quantum Leap, The New Physics for the Non-Scientists, Harper and Row Publishers, New York, NY